CHEMISTRY RESEARCH AND APPLICATIONS

EMULSIFIERS

PROPERTIES, FUNCTIONS AND APPLICATIONS

CHEMISTRY RESEARCH AND APPLICATIONS

Additional books in this series can be found on Nova's website under the Series tab.

Additional e-books in this series can be found on Nova's website under the e-book tab.

CHEMISTRY RESEARCH AND APPLICATIONS

EMULSIFIERS

PROPERTIES, FUNCTIONS AND APPLICATIONS

ADRIENNE FITZGERALD
EDITOR

nova publishers
New York

Library of Congress Cataloging-in-Publication Data

Emulsifiers : properties, functions, and applications / Adrienne Fitzgerald, editor.
pages cm. -- (Chemistry research and applications)
Includes bibliographical references and index.
ISBN 978-1-63483-688-3 (hardcover)
1. Emulsions. 2. Lipids in human nutrition. 3. Fatty acids. I. Fitzgerald, Adrienne.
TP156.E6E56 2015
660'.294514--dc23

2015031079

Published by Nova Science Publishers, Inc. † New York

CONTENTS

PREFACE

This book focuses on two kinds of emulsifiers. In the first chapter, surfactant and antioxidant properties of fatty acid esters synthesized through lipase-catalyzed condensation with various hydrophilic compounds is explored. In the second chapter, the impact of combined emulsifier on crystallization properties of non-trans fat is discussed. The third chapter provides a brief account of emulsifiers/stabilizers and their role in stabilizing complex colloid systems such as foamed emulsions, structured emulsions and bigels with the help of illustrative examples. The last chapter of the book explores lecithin, modified lecithins, polyglycerol polyricinioleate and sorbitan monostearate emulsifiers widely used in the food industry.

Chapter 1 – Fatty acid eaters were synthesized through condensation with various hydrophilic compounds, such as monosaccharide, phenolic glycoside, ascorbic acid and erythorbic acid, using an immobilized lipase. As these esters would be edible due to edibility of each substrate for the condensation and to enzymatic synthesis by a lipase, they would be promising emulsifiers for food. In addition, a part of the esters have antioxidative ability. Therefore, the surfactant and antioxidant properties of the synthesized eaters were examined. Acyl mannoses with the saturated acyl chain lengths of 8 to 16 were continuously produced using a plug-flow-type reactor. The conversion of *ca.* 40% was maintained for at least 16 days and the productivity was estimated to be 350 g/L-reactor·day. The surface tensions of aqueous solutions of the produced acyl mannoses were measured at various concentrations, and the critical micelle concentration, CMC, and the residual area per molecule, *a*, were calculated. The longer the acyl chain length was, the CMC was lower, while the *a* value scarcely depended on the acyl chain length. As acyl mannose molecules would be oriented so as to stick their acyl residues out in the air, the *a* value seemed to be exclusively determined by the mannose moiety.

The antimicrobial activity of myristoyl, palmitoyl, or stearoyl hexose, which was glucose, mannose, or galactose, coexistent with lysozyme against Gram-positive bacteria was measured to investigate the availability of acyl hexose as an antimicrobial co-agent. The stearoyl hexose coexistent with the lysozyme showed the highest activity against *Bacillus subtilis* and *Bacillus licheniformis* and could exhibit a higher activity than only the lysozyme. It was indicated that the antimicrobial action of the acyl hexose would be exerted parallel with the bacterial lysis of lysozyme.

Three lauroyl phenolic glycosides were synthesized through the condensation of phenolic glycoside, such as arbutin, naringin and phloridzin, with lauric acid. The suppressive ability of each lauroyl phenolic glycoside against the oxidation of linoleic acid was higher than that of the corresponding phenolic glycoside, whereas there was no difference between the radical scavenging activities of unmodified and lauroyl phenolic glycosides.

Acylation of ascorbic and erythorbic acids with a fatty acid also significantly improved the oxidative stabilities of linoleic acid and methyl linoleate. The kinetic equation of the autocatalytic type was applied to the oxidation processes of linoleic acid mixed with acyl ascorbates. The rate constant, k, value for the oxidation with the ascorbate was lower than that with no additive and ascorbic acid at any tested temperature, and there was little difference among the k values with the ascorbates having different acyl chains.

On the other hand, the Y_0 values, which were the initial fraction of unoxidized linoleic acid, with the ascorbates were greater than that with no additive. These results indicated that the addition of acyl ascorbate delayed the induction period in the oxidation process of linoleic acid. Accordingly, these fatty acid esters could be considered to be a useful food additive as an antioxidative emulsifier.

Chapter 2 – This research examined the influence of combination of two kind of emulsifiers and combined emulsifier 2 in 1 on crystallization properties of non trans fat. Nuclear magnetic resonance (NMR) spectroscopy was used for measuring the solid fat content (SFC) of fat samples at different temperatures, as well as for crystallization rate under static conditions, by measuring the change of SFC as a function of time. Kinetics of crystallization was defined applying Gompertz's mathematical method. Crystallization behavior of fat was also monitored using rotational viscometry and texture analyzer, while the melting point of fat was determined by differential scanning calorimetry (DSC).

Rheological measurements and crystallization kinetics havc showed that both types of emulsifiers accelerated the crystallization in relation to fat without emulsifiers. Fat samples with emulsifier 2 in 1 had a higher crystallization rate and thus formation of lower amount of smaller crystalls, compared to fat samples with combination of two emulsifiers.

Texture analyses also showed better spreadability of those samples in relation to samples with combination of two emulsifiers. On the other hand, the presence of emulsifiers in fat increased the melting point and melting enthalpy, especially the addition of combined emulsifier 2 in 1.

Chapter 3 – Colloidal systems such as foams, emulsions, gels and hybrid colloids such as emulsions gels and foamed emulsions are routinely encountered when dealing with food formulations. To ensure short and long term stability of these systems, a range of small molecular weight surfactants (including phospholipids, mono and di-acylglycerols etc.) and/ or polymeric stabilizers (proteins and modified polysaccharides) are commonly used as formulation aids. In both cases, the interfaces are stabilized by adsorption of molecular layers. While, small molecular weight surfactants have higher surface activity (i.e., they lower interfacial tension at low concentrations), the macromolecular stabilizers have a better adsorption efficiency (due to their large sizes, they are able to cover the interfaces efficiently, leading to the formation of stable viscoelastic interfacial layers). Recently, a third category of emulsifiers/stabilizers have generated a lot of interest where the interfaces are stabilized by formation of an adsorbed layer of solid particles in the colloidal size range. In this chapter, a brief account of such particles (including colloidal complexes, fine crystallites and inorganic particles) and their role in stabilizing complex colloid systems such as foamed emulsions, structured emulsions and bigels is discussed with the help of illustrative examples.

Chapter 4 – Lecithin, modified lecithins, polyglycerol polyricinoleate and sorbitan monostearate are emulsifiers widely used in food industry, with intrinsic abilities to modify some specific characteristic of lipid systems. In chocolate industries, standard soy lecithin and polyglycerol polyricinoleate (PGPR) are commonly added to the chocolate mass mainly for rheological adequacy. These two additives have complementary effects on viscosity and yield stress. Lecithin is reported to have greater effect on the plastic viscosity and PGPR reduces the yield stress. There are several types of lecithins and modified soy lecithins emerging as alternatives to expand the range of applications of phospholipids containing emulsifiers. The modified lecithins are produced by chemical, enzymatic or physical modification of standard lecithins. These changes affect the balance between hydrophilic and lipophilic group (HLB) of emulsifiers molecules, and hence, different effects are expected on food products. In addition, sorbitan monostearate, that is produced by esterification of sorbitol with a stearic acid, can be used as an active agent to improve the consistency and the heat resistance of oil and fat blends, currently applied for the development of low saturated lipid materials. Besides changing structural and rheological properties in food formulations, these emulsifiers also present the ability of altering lipid crystallization kinetics, and therefore, they are commonly termed as crystallization modifiers of fats. Determined by their chemical structure, emulsifiers exhibit an important role in the physical characteristics of fat-based products by delaying or speeding up the crystal nucleation, and even by changing morphology and packing density of the crystal network in the lipid matrix. Standard soy lecithins can act as crystallization promoters or inhibitors, depending on the concentration added to the fat phase. The addition of PGPR and standard soy lecithin to cocoa butter produces numerous crystals of small sizes and few crystals of larger size, respectively. Acyl–acyl interactions between additives and lipid systems are improved by similar chain length present in the fatty acids. Considering emulsifiers with structural similarities to triacylglycerol molecules, one accepted action mechanisms for explaining the fat crystals network modification is by co-crystallization. Other alternative mechanisms are considered for elucidating the variations in the crystal nucleation and growth after emulsifier additions. Sorbitan monostearate functioning in lipids is attributed to a very specific crystal network formation by the self-assemble capability of its structure. In this context, this chapter presents a comprehensive review on the structural characteristics and modification effects in lipid systems given by the addition of lecithins, polyglycerol polyricinoleate and sorbitan monostearate added to cocoa butter, palm oil and other fat blends.

In: Emulsifiers
Editor: Adrienne Fitzgerald

ISBN: 978-1-63483-688-3

Chapter 1

SURFACTANT AND ANTIOXIDANT PROPERTIES OF FATTY ACID ESTERS SYNTHESIZED THROUGH LIPASE-CATALYZED CONDENSATION WITH VARIOUS HYDROPHILIC COMPOUNDS

Yoshiyuki Watanabe[1,*] ***and Shuji Adachi***[2]

[1]Department of Biotechnology and Chemistry, Faculty of Engineering, Kinki University, Takaya, Higashi-Hiroshima, Japan

[2]Division of Food Science and Biotechnology, Graduate School of Agriculture, Kyoto University, Sakyo-ku, Kyoto, Japan

ABSTRACT

Fatty acid eaters were synthesized through condensation with various hydrophilic compounds, such as monosaccharide, phenolic glycoside, ascorbic acid and erythorbic acid, using an immobilized lipase. As these esters would be edible due to edibility of each substrate for the condensation and to enzymatic synthesis by a lipase, they would be promising emulsifiers for food. In addition, a part of the esters have antioxidative ability. Therefore, the surfactant and antioxidant properties

---

[*] E-mail address: wysyk@hiro.kindai.ac.jp

of the synthesized eaters were examined. Acyl mannoses with the saturated acyl chain lengths of 8 to 16 were continuously produced using a plug-flow-type reactor. The conversion of *ca.* 40% was maintained for at least 16 days and the productivity was estimated to be 350 g/L-reactor· day. The surface tensions of aqueous solutions of the produced acyl mannoses were measured at various concentrations, and the critical micelle concentration, CMC, and the residual area per molecule, *a*, were calculated. The longer the acyl chain length was, the CMC was lower, while the *a* value scarcely depended on the acyl chain length. As acyl mannose molecules would be oriented so as to stick their acyl residues out in the air, the *a* value seemed to be exclusively determined by the mannose moiety.

The antimicrobial activity of myristoyl, palmitoyl, or stearoyl hexose, which was glucose, mannose, or galactose, coexistent with lysozyme against Gram-positive bacteria was measured to investigate the availability of acyl hexose as an antimicrobial co-agent. The stearoyl hexose coexistent with the lysozyme showed the highest activity against *Bacillus subtilis* and *Bacillus licheniformis* and could exhibit a higher activity than only the lysozyme. It was indicated that the antimicrobial action of the acyl hexose would be exerted parallel with the bacterial lysis of lysozyme.

Three lauroyl phenolic glycosides were synthesized through the condensation of phenolic glycoside, such as arbutin, naringin and phloridzin, with lauric acid. The suppressive ability of each lauroyl phenolic glycoside against the oxidation of linoleic acid was higher than that of the corresponding phenolic glycoside, whereas there was no difference between the radical scavenging activities of unmodified and lauroyl phenolic glycosides.

Acylation of ascorbic and erythorbic acids with a fatty acid also significantly improved the oxidative stabilities of linoleic acid and methyl linoleate. The kinetic equation of the autocatalytic type was applied to the oxidation processes of linoleic acid mixed with acyl ascorbates. The rate constant, k, value for the oxidation with the ascorbate was lower than that with no additive and ascorbic acid at any tested temperature, and there was little difference among the k values with the ascorbates having different acyl chains.

On the other hand, the Y_0 values, which were the initial fraction of unoxidized linoleic acid, with the ascorbates were greater than that with no additive. These results indicated that the addition of acyl ascorbate delayed the induction period in the oxidation process of linoleic acid. Accordingly, these fatty acid esters could be considered to be a useful food additive as an antioxidative emulsifier.

Keywords: Antimicrobial activity, antioxidant, enzymatic synthesis, fatty acid ester, surfactant

1. INTRODUCTION

Acyl saccharides, which are products from the condensation of a fatty acid with the mono- or disaccharide, are biosurfactants with good emulsifying properties [1-3] and are of much interest for use in the food, cosmetics, and pharmaceuticals industries [4].

They have been produced on an industrial scale based on chemical procedures.

Their syntheses through lipase-catalyzed transesterification [5-9] or condensation [10-24] reaction would have some advantages; i.e., high regioselectivity of the enzyme, the moderate reaction conditions and the direct use of unmodified substrates. Fatty acid eaters were synthesized through condensation with various hydrophilic compounds using an immobilized lipase.

As these esters would be edible due to edibility of each substrate for the condensation and due to enzymatic synthesis by a lipase, they would be promising emulsifiers for food. Furthermore, a part of the esters have antioxidative ability.

Therefore, the surfactant and antioxidant properties of the synthesized eaters were examined. In next section, a continuous production of 6-*O*-acyl mannose through the immobilized-lipase-catalyzed condensation of saturated fatty acid and mannose using a plug-flow-type reactor was examined [25]. The surface tensions in aqueous solution of the products were measured and their surfactant properties were evaluated. In Section 3, the antimicrobial activities of the acyl hexose, which was synthesized through the condensation of myristic, palmitic or stearic acid with glucose, mannose or galactose, coexisting with the lysozyme against *Bacillus coagulans*, *Bacillus subtilis* and *Bacillus licheniformis* were examined [26].

In Section 4, three lauroyl phenolic glycosides were synthesized by using arbutin, naringin or phloridzin and the antioxidative activities of lauroyl phenolic glycosides against lipid oxidation were compared each other [27]. In addition, the antioxidative property of acyl erythorbate for lipid oxidation was also investigated [28].

D-Erythorbic acid is a stereoisomer of L-ascorbic acid. At last, the oxidation processes of linoleic acid in the presence of ascorbic acid or saturated acyl ascorbate were measured at the various molar ratios and the kinetically analyze was executed by the rate expression of autocatalytic type [29].

2. Continuous Production of Fatty Acid Esters with Mannose by Immobilized Lipase and Their Surfactant Properties

Condensation of saturated fatty acids and mannose was continuously carried out using a plug-flow-type reactor with an immobilized lipase and the surfactant properties of the products were evaluated [25]. Immobilized lipase from *Candida antarctica*, Chirazyme® L-2 c.-f. C2 from Roche Molecular Biochemicals, Mannheim, Germany, were packed into a cylindrical glass column (11.5 mmϕ × 150 mm). Mannose and a fatty acid (octanoic, decanoic, lauric, myristic or palmitic acid) were dissolved with the dehydrated 2-methyl-2-propanol. The substrate solution was fed to the column at a specified flow rate by a delivery pump. The substrate reservoir, column and pump were installed in a thermo-regulated chamber at 50°C. The product concentration was determined using an HPLC. The operations for continuous production were carried out at various flow rates for every fatty acid to obtain the relationship between the conversion and the superficial residence time, τ_0. A mixture of 20 mmol/L mannose and 100 mmol/L myristic acid solutions was continuously fed to the column packed with the immobilized lipase (0.75 g by dry weight, 10 mmϕ × 50 mm). The substrate solution was fed to the column at various flow rates and the conversion of mannose to acyl mannose at a steady-state was measured. The conversion was defined as a ratio of the product concentration in the effluent to the mannose one in the feed solution. For every fatty acid, the conversion of more than 0.5 was attained at $\tau_0 \geqq 20$ min. There was a weak tendency that the equilibrium conversion was higher for the mannose ester with the longer acyl chain. The condensation product between mannose and lauric acid was analyzed by ^{1}H NMR and identified to be 6-*O*-lauroyl mannose. The long-term operational stability of the enzyme was examined for the synthesis of the myristoyl, lauroyl and decanoyl mannoses at τ_0 = 12 min. The substrate solution was fed for 16 days at a flow rate of 0.33 mL/min, which corresponded to τ_0 = 12 min. On day 16, the feed solution was changed to a mixture of mannose and lauric acid, the concentrations of which were 20 mmol/L and 100 mmol/L, respectively, and the production of lauroyl mannose was continued for 3 days. In addition, the decanoyl mannose was then continuously produced under similar conditions for 3 days. As shown in Figure 1, myristoyl mannose was produced at a constant conversion of *ca.* 0.4 for at least 16 days. The productivity during the operation was evaluated to be 350 g/L-reactor·day. The lauroyl and decanoyl mannoses were also produced at similar conversions. The effluent was rotary-evaporated to reduce its volume to about half.

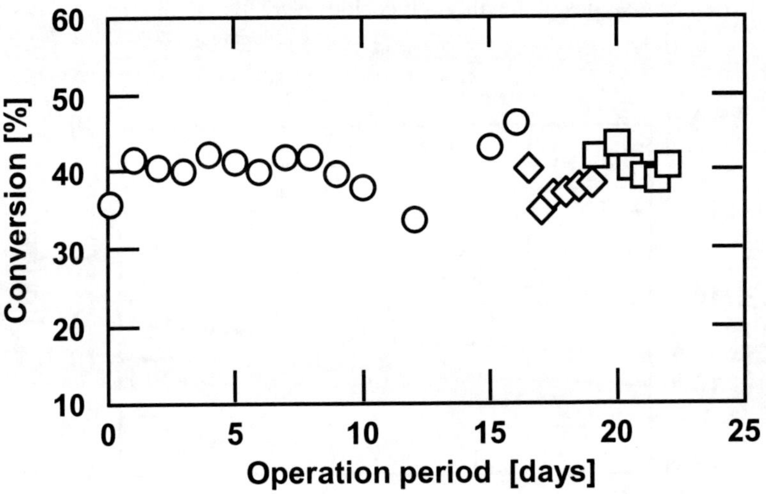

Figure 1. Continuous production of (○) myristoyl, (◇) lauroyl and (□) decanoyl mannoses by a plug-flow-type reactor of immobilized lipase at $\tau_0 = 12$ min.

The concentrated effluent was applied to an ODS column (20 mmϕ × 250 mm) and eluted with a mixture of acetonitrile and water (65:35, by vol.) at a flow rate of 7 mL/min. The effluent at the peak corresponding to the desired product, which was monitored with a refractometer, was collected, and the product was recovered by evaporation. The product was dissolved with water at various concentrations and the surface tension was measured by the Wilhelmy method using a surface tensiometer at 25°C. The critical micelle concentration, CMC, and the surface excess, Γ, was estimated from the measured surface tension and the following equation:

$$-\frac{d\gamma}{d\log C} = \frac{RT}{0.434}\Gamma \tag{1}$$

where γ is the surface tension, C is the concentration of the acyl mannose, R is the gas constant and T is the absolute temperature. The reciprocal of the Γ value gives the residual area per molecule, a. These surfactant properties of the acyl mannoses are shown in Figure 2. The longer acyl chain was, the lower CMC was. On the other hand, the surface tension at critical micelle concentration, γ_{CMC}, the Γ and a values scarcely depended on the acyl chain length.

The a values were practically same among the acyl mannoses and were ca. 0.40 nm^2. The a values of the alkyl glycosides were also in the range of 0.37 to 0.49 nm^2 [30].

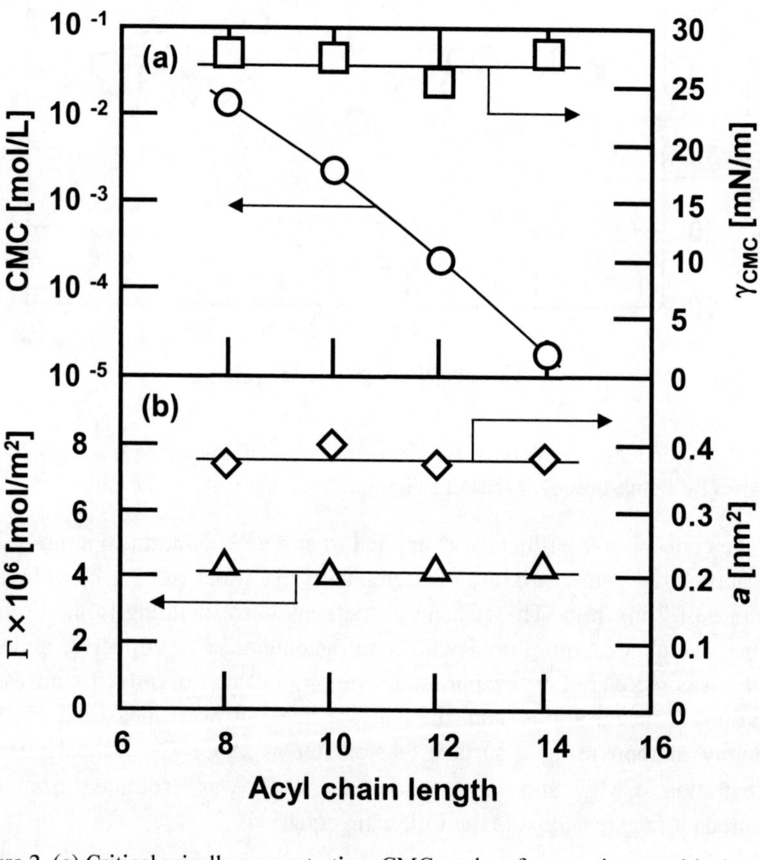

Figure 2. (a) Critical micelle concentration, CMC, and surface tension at critical micelle concentration, γ_{CMC}, and (b) surface excess, Γ, and residual area per molecule, a, of acyl mannoses at 25°C.

As acyl mannose molecules would be oriented so as to stick their acyl residues into the air, the *a* value seemed to be exclusively determined by the saccharide moiety.

3. Antimicrobial Activity of Fatty Acid Esters with Hexose against Gram-Positive Bacilli

Many surfactants have an antimicrobial ability. Acyl glycerols and sucroses are commonly used to prevent or suppress bacterial spore development in canned soft drinks stored at 50 to 70°C in the cold and cool seasons [31, 32]. The bacteriostatic activities of monoacyl sugar alcohols with different acyl chains and hydrophilic heads have been reported against some thermophilic sporeformers [33]. Lysozyme (EC 3.2.1.17) catalyzes the hydrolytic reaction for the β(1-4) bond between *N*-acetylmuramic acid and *N*-acetylglucosamine in polysaccharides present in the cell wall of bacteria, and thus is used as an antimicrobial agent due to its bacterial lytic ability [34]. The lytic activity of lysozyme, however, decreases during the heating-process when manufacturing foods due to inactivation by heating. Therefore, the addition of co-agents would be useful for suppressing the growth of microorganism by lysozyme in foods. The antimicrobial activities of the esters, which were synthesized through the lipase-catalyzed condensation of glucose, mannose or galactose with myristic, palmitic or stearic acid, coexisting with the lysozyme against several Gram-positive bacteria were evaluated in order to investigate the availability of acyl hexose as an antimicrobial co-agent [26]. *Bacillus coagulans* (NBRC3557), *Bacillus subtilis* (NBRC3007) and *Bacillus licheniformis* (NBRC12107) were tested as representative Gram-positive bacteria in foods. The syntheses of the monoacyl hexoses were carried out by using glucose, mannose or galactose (7.5 mmol) and octanoic, decanoic, lauric, myristic, palmitic or stearic acid (37.5 mmol), and an immobilized lipase from *C. antarctica*. The liquid medium was prepared by mixing peptone, yeast extract, magnesium sulfate heptahydrate and distilled water. The pH of the mixture was adjusted to 7.0 by the addition of sodium hydroxide. A freeze-dried type culture of *B. coagulans* was rehydrated with the autoclaved medium, and the inoculum was incubated by slowly shaking in a water-bath under anaerobic condition at 37°C. After dimethylsulfoxide containing a specific amount of acyl hexose was added to the prepared culture, the culture was incubated at 37°C under anaerobic condition. At appropriate intervals, the culture was sampled, and the optical density, OD, at 600 nm of the culture was measured using a spectrophotometer. The OD of the culture without acyl hexose was evaluated as the control, ODcontrol. The ratio of the OD of the culture with acyl hexose to ODcontrol was used as an index for the antimicrobial activity. The antimicrobial activity of acyl hexose coexistent with lysozyme using the disc diffusion test was analyzed using the reported methods [35, 36]. *B. subtilis* or *B. licheniformis* was inoculated into the autoclaved liquid medium composed of peptone, sodium chloride, agar, meat extract and magnesium sulfate heptahydrate, then heat-shocked at 80°C for 10 min. The medium was placed on an agar plate, and cultivated at 30°C for 15 h. The Gram-positive bacteria were then cultivated in the liquid medium at about 10^4 to 10^6 CFU. The culture was added to the agar medium, and solidified at room temperature. Equal weights of the acyl hexose to lysozyme were dissolved in dimethylsulfoxide at concentration from 10 mg/L to 1 g/L. The filter paper disc containing the solution was placed on the agar plate. The bacteria were cultivated at 37°C for 24 h, and the diameter of the zone of growth inhibition was measured.

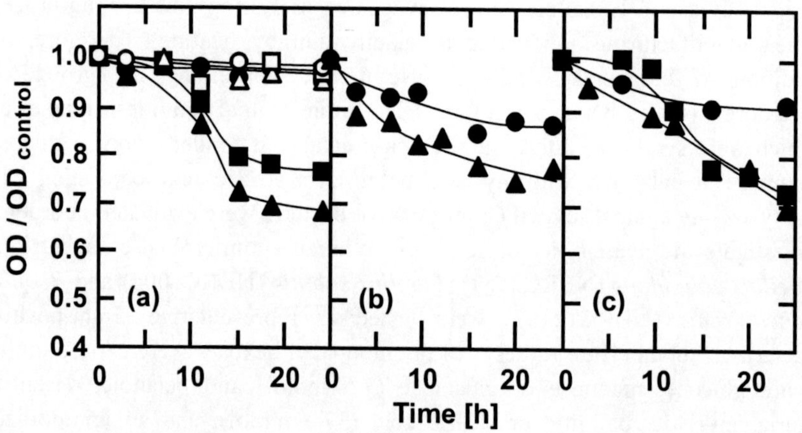

Figure 3. The growth inhibition of *B. coagulans* by (a) glucose, (b) mannose, and (c) galactose esters with (○) octanoic, (□) decanoic, (△) lauric, (●) myristic, (■) palmitic and (▲) stearic acids at the concentration of 40 mg/L and 37°C.

Figure 3 shows the growth inhibition of *B. coagulans* by monoacyl glucose, mannose, and galactose with various acyl chain lengths from 8 to 18 at the concentration of 40 mg/L. The ratio of the OD for the culture with octanoyl, decanoyl, lauroyl or myristoyl glucose to the ODcontrol did not change with time as shown in Figure 3(a), whereas the palmitoyl and stearoyl glucoses exhibited an antimicrobial activity after 15 h. The myristoyl and stearoyl mannoses also inhibited the growth of *B. coagulans*, though these esters gradually suppressed the growth from the beginning (Figure 3(b)). Myristoyl galactose slowly inhibited the growth, but the antimicrobial activity was low (Figure 3(c)). The OD/ODcontrol value for the culture with palmitoyl or stearoyl galactose more rapidly decreased than that with the myristoyl ester. Many studies indicate that the dependency of the antimicrobial activity of the acyl saccharides on the acyl chain length and the saccharide moiety was different between the microorganism species [37-41].

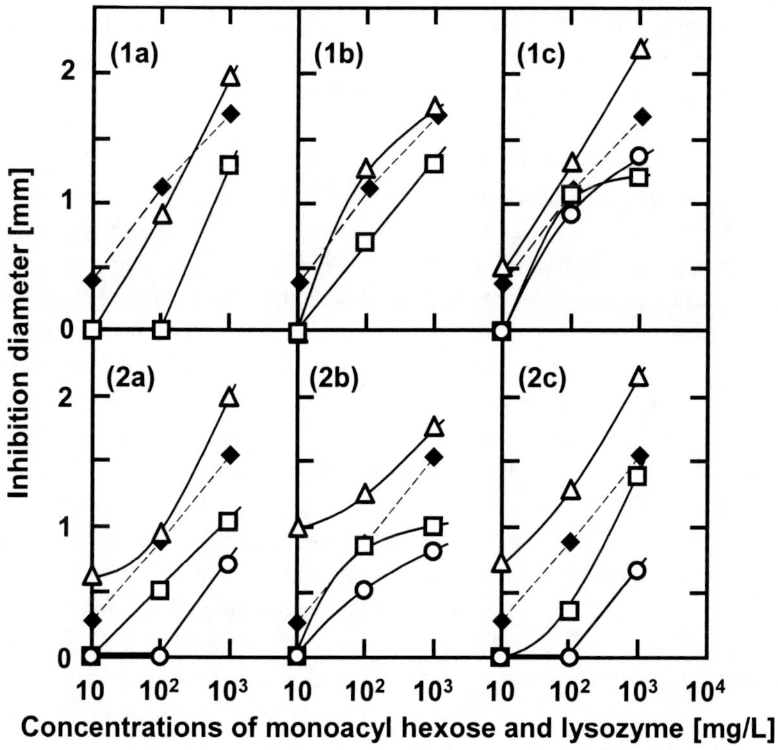

Figure 4. Effect of the concentrations of (a) glucose, (b) mannose and (c) galactose esters with (○) myristic, (□) palmitic and (△) stearic acids, and lysozyme on inhibition diameters in disc diffusion test for (1) *B. subtilis* and (2) *B. licheniformis*. Closed square symbols (◆) represent the sole addition of lysozyme.

The results in Figure 3 also showed that the antimicrobial activity of the acyl hexose depended on its hydrophobicity, though the mechanism of the antimicrobial action of the acyl hexose remains unclear. Moriyama et al. reported that the antimicrobial activity of acyl sucrose was related to its adsorption on the spore coat proteins of Bacillus cereus [42]. The dependence of the antimicrobial activity of acyl hexose on its hydrophobicity would be consistent with the results from its adsorption on bacterial spore-coating proteins. The antimicrobial activity of myristoyl, palmitoyl or stearoyl hexose coexistent with lysozyme against the two Gram-positive bacteria was measured. Figure 4 show the effect of the concentrations of acyl glucose, mannose or galactose, and lysozyme on the inhibition diameters in the disc diffusion test for *B. subtilis* and *B. licheniformis*, respectively. As shown in Figure 4(1), the antimicrobial activity of stearoyl hexose coexisting with the lysozyme against *B. subtilis* was the highest in every hexose ester. Stearoyl hexose coexisting with the lysozyme at 1,000 mg/L exhibited a higher antimicrobial activity than the lysozyme alone. The antimicrobial activities of 100 mg/L stearoyl mannose and 10 and 100 mg/L galactose were slightly higher than that of the lysozyme, but each stearoyl hexose hardly suppressed the growth of *B. subtilis* at 10 mg/L. The antimicrobial activities of the myristoyl and palmitoyl hexoses coexisting with the lysozyme were lower than that of the lysozyme alone at all the concentrations. It was found that stearoyl hexose at 1,000 mg/L was effective as a co-agent of the lysozyme for the antimicrobial activity against *B. subtilis*, but the cooperative antimicrobial activity of the monoacyl hexose with the lysozyme was generally low. It was also shown in Figure 4(2) that stearoyl hexose coexisting with the lysozyme exhibited the highest antimicrobial activity against *B. licheniformis*. The activities of the myristoyl and palmitoyl hexoses coexisting with the lysozyme were lower than that of the lysozyme. Compared to the results for *B. subtilis*, stearoyl glucose and mannose with the lysozyme showed a higher activity against *B. licheniformis* than the lysozyme at both 10 and 100 mg/L. Stearoyl galactose inhibited the growth of both bacteria at all the concentrations. The lysozyme is endo-β-1, 4-*N*-acetylhexosaminidase and catalyzes the hydrolysis of the β-1, 4-bond between *N*-acetylglucosamine and *N*-acetylmuramic acid in the cell wall, resulting in bacterial lysis [43]. The reason why the antimicrobial activities of some acyl hexoses coexisting with the lysozyme were lower than that of the lysozyme would be due to the inhibition of the lytic action of the lysozyme by the acyl hexose. If the antimicrobial action of the acyl hexose is due to its adsorption on spore proteins, this is parallel with the lytic action of lysozyme. Therefore, stearoyl hexose, which had a high hydrophobicity and adsorption ability on spores, coexisting with the lysozyme could exhibit a higher antimicrobial activity than only the lysozyme. The high antimicrobial activity against *B. licheniformis* by the stearoyl hexose at low concentrations may be due to its high affinity to the spore proteins. Stearoyl hexose, especially stearoyl galactose, would be available for the inhibition of some Gram-positive bacteria as an antimicrobial co-agent.

4. Suppressive Ability of Fatty Acid Esters with Various Hydrophilic Compounds against Lipid Oxidation

Phenolic glycosides from crude plant materials are increasingly receiving attention in relation to their antioxidative activities [44-46]. Arbutin found in plant *Uvae ursi* is used in cosmetics due to its whitening effect on the skin [47, 48]. Naringin is the primary bitter component in citrus fruits such as grapefruit but the bitterness in citrus fruit juices is one of the major problems of the citrus industry [49, 50]. Phloridzin is included in apple skin and has the lowering effect on the postprandial blood glucose level *in vivo* [51]. These phenolic glycosides would have also antioxidative activity due to their phenolic hydroxyl groups. Three lauroyl phenolic glycosides through the condensation of arbutin, naringin or phloridzin with lauric acid, which was used for syntheses of various fatty acid esters with high surface activity as a representative fatty acid, were synthesized by an immobilized lipase from *C. antarctica* in various organic solvents [27]. In addition, the antioxidative activity of each lauroyl phenolic glycoside against the oxidation of linoleic acid and 1,1-diphenyl-2-picrylhydrazyl (DPPH) free radical scavenging activity were also measured. The oxidation process of linoleic acid with lauroyl or unmodified phenolic glycoside was measured as follows: Linoleic acid was dissolved in hexane for the solution. Each lauroyl or unmodified phenolic glycoside dissolved in ethanol was added to the linoleic acid solution at the molar ratio of 0.1 to linoleic acid. The mixture was placed in flat-bottomed glass cups, and the hexane and ethanol was then evaporated under reduced pressure in a desiccator. The cups were placed in a plastic container in which a Petri dish filled with saturated lithium chloride aqueous solution to maintain the relative humidity at 12%. The container was then stored in the dark at 50°C. The cups were periodically taken out of the desiccator, and a mixture of methanol, benzene and methyl myristate (20:80:0.05, by vol.) was added to the cup. Linoleic acid was converted to its methyl ester by adding trimethylsilyldiazomethane solution dissolved in hexane and allowing it to stand at room temperature for 30 min [52]. After evaporation under reduced pressure, the remainder was dissolved in hexane, and the solution was used for the GC analysis. The amount of unoxidized linoleic acid was determined by a gas chromatograph with a hydrogen ionization detector and a capillary column, the dimensions of which were 0.25 mm in diameter and 30 m in length, with polyethylene glycol. The injection or detection temperature was 200ºC, and column temperature was 180ºC. Based on a reported method [53], the radical scavenging activities of unmodified and lauroyl phenolic glycoside were measured. A 50% ethanol solution of lauroyl or unmodified phenolic glycoside (0.125 mmol/L) and 0.5 mmol/L ethanol solution of DPPH radical were added to an amber vial. The headspace of the vial was filled with nitrogen gas and it was tightly sealed. The vial was then vigorously shaken and incubated for 20 min at 25ºC. The radical scavenging activity of each phenolic glycoside was measured by the decolorization of DPPH radical at 516 nm using a UV-VIS spectrophotometer. Figure 5 shows the oxidation processes of linoleic acid with lauroyl or unmodified phenolic glycoside. The fraction of unoxidized linoleic acid without lauroyl and unmodified phenolic glycoside rapidly decreased as shown in Figure 5(a), whereas Figs. 5(b)-(d) show that the oxidation with lauroyl or unmodified phenolic glycoside slowly proceeded. Furthermore, the suppressive effect of each lauroyl phenolic glycoside against the oxidation was higher than those of the corresponding phenolic glycoside. The oxidation kinetics was empirically expressed by the Weibull equation, which is flexible and has a potential for describing many deterioration kinetics [54]:

$$Y = \exp[-(kx)^n] \tag{2}$$

where Y is the fraction of unoxidized linoleic acid at time t, k is the rate constant and n is the shape constant. The kinetic parameters, k and n, were evaluated by fitting the experimental results by nonlinear regression. The curves in Figure 5 were drawn based on the equation using the estimated parameters. Figure 6 shows the comparison between the k values estimated for the oxidation with lauroyl and unmodified phenolic glycoside. It was quantifically shown that the antioxidative activity of each lauroyl phenolic glycoside against the oxidation was higher than was higher than that of unmodified phenolic glycoside, as all symbols were under the diagonal line in Figure 6. The length from the position of the symbol to the diagonal would reflect the degree of the antioxidative effect for acylation of phenolic glycoside against the oxidation. The n values for the oxidation with lauroyl and unmodified phenolic glycoside are 3.12 for lauroyl arbutin, 1.38 for lauroyl naringin, and 1.34 for lauroyl phloridzin, 0.787 for arbutin, 2.92 for naringin, 2.52 for phloridzin, respectively. There was no tendency between the n values for the oxidation with lauroyl and unmodified phenolic glycoside. The Weibull model has characteristics of a sigmoidal pattern that can be described when $n > 1$, that the model expresses the first-order kinetics at $n = 1$, and that Y value steeply decreases during the early stage when $n < 1$. Therefore, it was indicated that each oxidation of linoleic acid except that with arbutin was represented as a sigmoidal pattern.

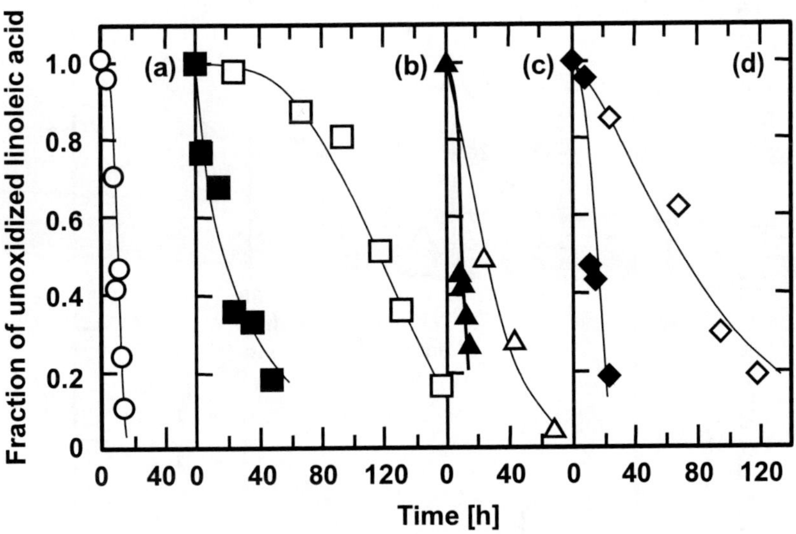

Figure 5. Oxidation stabilities of linoleic acid with (○) no additive, (□, ■) arbutin, (△, ▲) naringin and (◇, ◆) phloridzin at 50°C and 12% relative humidity. The closed and open symbols represent unmodified and acyl polyphenol glycoside, respectively. The solid curves were calculated using the estimated kinetic parameters of the Weibull equation.

The inset in Figure 6 shows DPPH radical scavenging activities of lauroyl and unmodified phenolic glycoside. Each symbol was on the diagonal line in the inset, indicating that there was no difference between the radical scavenging activities of lauroyl and unmodified phenolic glycosides in ethanol solution. The radical scavenging activity of lauroyl phenolic glycoside, however, depended on the type of phenolic glycoside, and the acitivity was high in order of lauroyl arbutin, phloridzin and naringin. This order was same as that for their antioxidative effect against the oxidation of linoleic acid. The antioxidative activity of the glycoside would depend on the number and the placement of phenolic hydroxyl group exhibiting the activity in the molecule.

L-Ascorbic acid, known as vitamin C, is a water-soluble vitamin and is widely used in foods as an antioxidant owing to its strong reducing ability. Ascorbic acid in some vegetables is easily oxidized through the catalysis of ascorbate oxidase, and the oxidation leads to a decrease in the antioxidative ability and bioavailability of ascorbic acid.

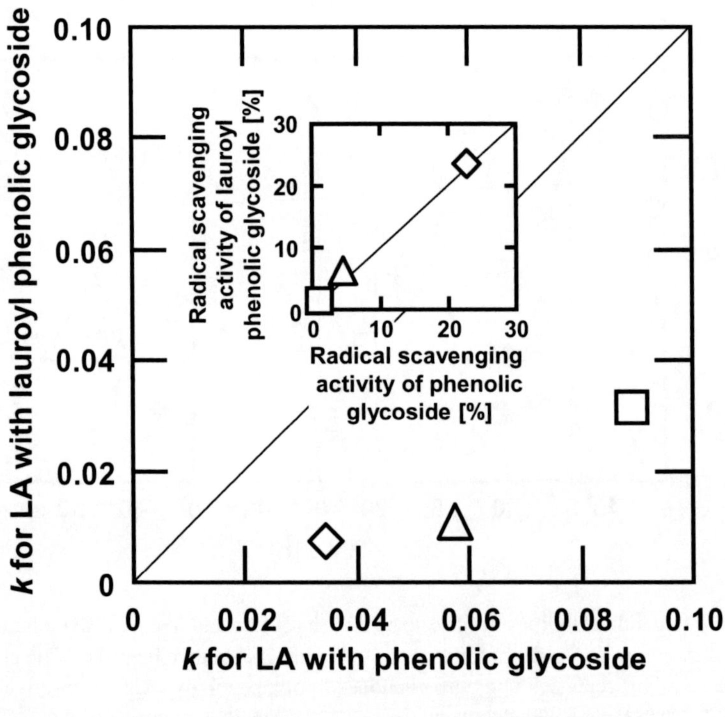

Figure 6. Comparison between the rate constants, *k*, estimated by the Weibull model for the oxidation of linoleic acid with acyl and unmodified (◇) arbutin, (□) naringin and (Δ) phloridzin. The inset shows DPPH radical scavenging activities of acyl and unmodified polyphenol glycoside at 20°C.

D-Erythorbic acid is a stereoisomer of ascorbic acid and is a by-product in the synthesis of ascorbic acid from glucose using both chemical and microbial processes [55]. Erythorbic acid has only *ca.* 5% of the vitamin activity of ascorbic acid [56], although it is approved as a food antioxidant due to its reducing properties. Obata et al. showed that erythorbic acid has a weaker affinity for ascorbate oxidase than ascorbic acid, and is far less prone to enzymatic oxidation [57]. Erythorbic acid would serve as a more effective antioxidant than ascorbic acid in vegetables containing ascorbate oxidase. The synthesis of 6-*O*-acyl ascorbate through the lipase-catalyzed condensation of a fatty acid and ascorbic acid has been reported [58, 59]. The enzymatic condensation of erythorbic acid and lauric acid has also been reported [60]. However, the antioxidative properties of the product, acyl erythorbate, for lipid oxidation have not been investigated. Erythorbate efficiently exerts antioxidative activity in a food system despite the presence of ascorbate oxidase. Acyl erythorbates were synthezed using the above-mentioned immobilized lipase, and the suppressive ability against lipid oxidation was evaluated [28]. The 50% DPPH radical scavenging concentrations, SC_{50}, were estimated. The SC_{50} values of octanoyl, decanoyl, lauroyl, myristoyl and palmitoyl erythorbate, erythorbic acid, palmitoyl ascorbate and ascorbic acid were 5.33, 5.39, 5.06, 5.19, 5.68, 5.41, 4.72 and 5.28 μmol/L, respectively. These data indicated that there was no difference in radical scavenging activity between acyl erythorbates and erythorbic acid and between erythorbate and ascorbate in ethanol solution. The peroxide value of methyl linoleate with acyl erythorbate was measured as follows: a plastic container was maintained relative humidity at 12%. The container was stored in the dark at 65°C for 1 day. Methyl linolate was dissolved in hexane, followed by palmitoyl erythorbate or erythorbic acid dissolved in ethanol. The palmitoyl erythorbate or erythorbic acid was added to methyl linolate at a molar ratio of 0.01. The mixture was placed in flat-bottomed glass cup, and hexane and ethanol were evaporated under reduced pressure in a desiccator. The cups were placed in the container and stored at 65°C. Each cup was periodically removed, and a mixture of chloroform and methanol (1:2, by vol.) was added to the sample in the cup. Moreover, a 25 mmol/L hydrochloric acid/methanol solution and the same volume of a 12.5 mmol/L ammonium/iron(III) sulfate solution were added, and the mixture was fully agitated by using the test tube mixer. Then, a saturated potassium iodide aqueous solution was added, and the sample was centrifuged.

The absorbance of the supernatant after the addition of the saturated potassium iodide solution was measured at 363 nm using the above-mentioned spectrophotometer. Figure 7 shows the transient changes in the peroxide value of methyl linoleate with erythorbates. The suppressive effect of erythorbic acid on the oxidation was very similar to that of ascorbic acid. Palmitoyl erythorbate and ascorbate are more effective at preventing oxidation than erythorbic and ascorbic acids. The suppressive abilities of the palmitoyl erythorbate and ascorbate against oxidation were similar, and it was consistent with the radical scavenging activity results. The enhancement of the solubilities of erythorbate and ascorbate in a bulky lipid by the introduction of the acyl group to erythorbic and ascorbic acids would contribute to the improvement of the suppressive ability against lipid oxidation.

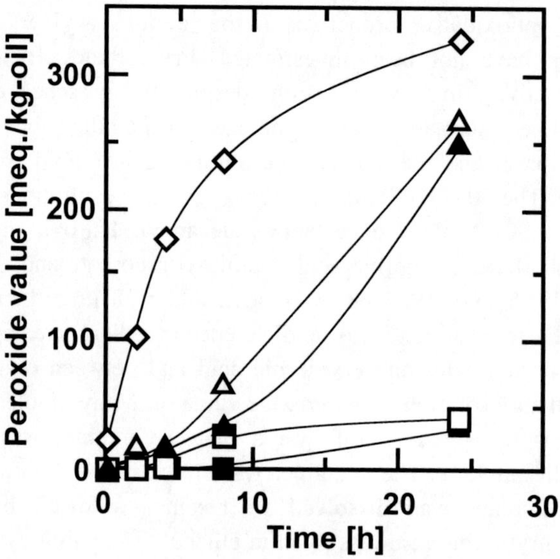

Figure 7. Changes in the peroxide value for the oxidation of methyl linoleate with (◇) no additive, (△) erythorbic acid, (□) palmitoyl erythorbate, (▲) ascorbic acid, and (■) palmitoyl ascorbate at 0.01 molar ratio of each additive to methyl linoleate. The oxidation were carried out at 65°C and under 12% relative humidity.

The oxidation processes of linoleic acid in the presence of saturated acyl ascorbate or ascorbic acid were measured at the various molar ratios [29]. The autoxidation processes of linoleic acid mixed with saturated acyl ascorbate or ascorbic acid at various molar ratios were measured as follows: At first, linoleic acid and octanoyl, lauroyl or palmitoyl ascorbate or ascorbic acid were dissolved in methanol at specific concentrations. The linoleic acid solution was placed in the amber glass vials. Then, acyl ascorbate or ascorbic acid solution was added to the vials to give the molar ratio of 0.05 (ascorbate/

linoleic acid). In addition, methanol was added to the samples of the molar ratios of 0.05. For the oxidation process of linoleic acid without any additive, methanol was solely added to the linoleic acid solution. The mixture was placed in flat-bottomed glass cups, and the methanol was then evaporated under reduced pressure. The cups were placed in a plastic container at 12% relative humidity. The container was stored in the dark at 37, 50, 65 and 80°C. Samples were periodically taken, and methyl myristate solution (methanol/benzene/methyl myristate = 20:80:0.05, by vol.) was added. Then, 2.0 mol/L trimethylsilyl diazomethane solution was poured into the cup to convert unoxidized linoleic acid to methyl linoleate [52].

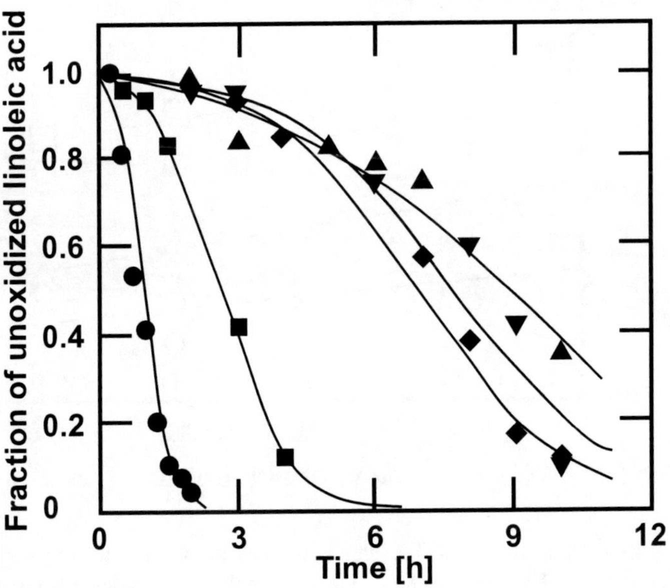

Figure 8. Oxidation processes of (●) linoleic acid with no additive and that mixed with ((■) ascorbic acid, (◆) octanoyl, (▼) lauroyl and (▲) palmitoyl ascorbate at the molar ratio = 0.05 and at 80°C. The solid curves were drawn using the k and Y_0 values estimated in the rate expression of the autocatalytic type.

After evaporation of methanol, benzene and hexane under reduced pressure, the remainder was then dissolved with hexane, which was used for the GC analysis with an FID [61]. Figure 8 shows the oxidation processes at 80°C and 12% RH of linoleic acid with no additive and that mixed with octanoyl, lauroyl, or palmitoyl ascorbate or ascorbic acid at the molar ratio (additive/linoleic acid) = 0.05. Linoleic acid with no additive was rapidly oxidized. When ascorbic acid was added to linoleic acid, the oxidative stability was slightly improved. Octanoyl, lauroyl, and palmitoyl ascorbates retarded the oxidation of linoleic acid more than ascorbic acid. There seemed to be little difference in the antioxidative ability among the three acyl ascorbates. The entire oxidation process of linoleic acid could be expressed by the following kinetic equation of the autocatalytic type [62, 63]:

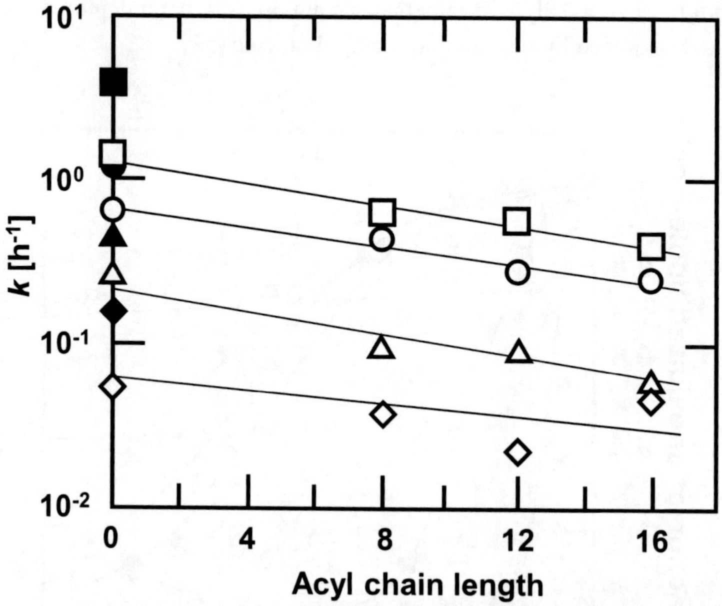

Figure 9. Relationship between acyl chain length of ascorbates and the rate constant, *k*, at (◇) 37, (△) 50, (○) 65 and (□) 80°C. Open symbols represent the rate constants for the oxidation of linoleic acid mixed with various ascorbates and closed symbols that for no additive.

$$\frac{dY}{dt} = -kY(1-Y) \tag{3}$$

Under the condition of $Y = Y_0$ at $t = 0$, the integration of Eq. 3 gives:

$$\ln\frac{1-Y}{Y} = kt + \ln\frac{1-Y_0}{Y_0} \tag{4}$$

where Y_0 is the initial fraction of unoxidized substrate and determines the induction period. Based on a linear regression analysis for linear plots of ln $[(1-Y)/Y]$ versus t for the oxidation process, the k and Y_0 values were determined from the slope and the intercept, respectively. The k and Y_0 values for the oxidation processes at 37, 50 and 65°C were also estimated. Figure 9 shows the relationship between the acyl chain length of the ascorbates and the k value at various temperatures. At any temperature, the k value with no additive was greater than that with acyl ascorbate or ascorbic acid. When octanoyl, lauroyl or palmitoyl ascorbate was added, there was a tendency for the k value in the presence of ascorbate with a larger acyl chain to be slightly lower.

The k value with ascorbic acid, the acyl chain length equal to zero, seemed to be greater than that with the acyl ascorbates at any temperature. As mentioned above, the Y_0 value reflects the initial state of the substrate, which affects the induction period, and the large Y_0 value indicates the elongation of the induction period.

The Y_0 values for the additives were greater than that with no additive, although there were some exceptions. This indicated that an additive delayed the induction period in the oxidation process of linoleic acid. The temperature dependence of the rate constant was analyzed based on the Arrhenius equation:

$$k = k_0 \exp(-\frac{E}{RT}) \tag{5}$$

where k_0 is the frequency factor, E is the apparent activation energy. The Arrhenius plots produced a straight line to evaluate the apparent activation energy, E, and the frequency factor, k_0, from the slope and the intercept of the line, respectively.

The E values for the oxidation of linoleic acid with every ascorbate were 50 to 70 kJ/mol, and there was a tendency for E value to decrease with increasing acyl chain length.

The k_0 values with ascorbates were lower than that with no additive or with ascorbic acid. The k_0 value also decreased with increasing acyl chain length. This would indicate that the presence of acyl ascorbate lowers both the height of the energy barrier for the oxidation of linoleic acid and the probability of the reaction resulting in the oxidation.

ACKNOWLEDGMENTS

The authors would like to thank all the staff of the Laboratories of Department of Biotechnology and Chemistry, Faculty of Engineering, Kinki University and Division of Food Science and Biotechnology, Graduate School of Agriculture, Kyoto University.

REFERENCES

[1] Sarney, D. B. & Vulfson, E. V. (1995). Application of enzymes to the synthesis of surfactants. *Trends in Biotechnology*, 13, 164-172.

[2] Scheckermann, C., Schlotterbeck, A., Schmidt, M., Wray, V. & Lang, S. (1995). Enzymatic monoacylation of fructose by two procedures. *Enzyme and Microbial Technology*, 17, 157-162.

[3] Ducret, A., Giroux, A., Trani, M. & Lortie, R. (1996). Characterization of enzymatically prepared biosurfactants. *Journal of American Oil Chemists' Society*, 73, 109-113.

[4] Nishikawa, Y., Yoshimoto, K., Nishijima, M., Fukuoka, F. & Ikekawa, T. (1987). Chemical and biological studies on carbohydrate esters. IX. Antitumor effects of selectively fatty acylated products of maltose. *Chemical and Pharmaceutical Bulletin*, 29, 505-513.

[5] Coulon, D., Girardin, M., Rovel, B. & Ghoul, M. (1995). Comparison of direct esterification and transesterification of fructose by *Candida antartica* lipase. *Biotechnology Letters*, 17, 183-186.

[6] Ferrer, M., Cruces, M. A., Bernbé, M., Ballesteros, A. & Plou, F. J. (1999). Lipase-catalyzed regioselective acylation of sucrose in two-solvent mixtures. *Biotechnology and Bioengineering*, 65, 10-16.

[7] Yan, Y., Bornscheuer, U., Cao, L. & Schmid, R. (1999). Lipase-catalyzed solid-phase synthesis of sugar fatty acid esters: Removal of byproducts by azeotropic distillation. *Enzyme and Microbial Technology*, 25, 725-728.

[8] Pérez-Victoria, I. & Morales J. C. (2007). One- and two-step enzymatic synthesis of polymerizable vinyladipoyl mono- and diesters of non-reducing trisaccharides. *Journal of Molecular Catalysis B: Enzymatic*, 48, 8-15.

[9] Walsh, M. K., Bombyk, R. A., Wagh, A., Bingham, A. & Berreau, L. M. (2009). Synthesis of lactose monolaurate as influence by various lipases and solvents. *Journal of Molecular Catalysis B: Enzymatic*, 60, 171-177.

[10] Khaled, N., Montet, D., Farines, M., Pina, M. & Graille, J. (1992). Synthesis of sugar mono-esters by biocatalysis. *Oléagineux*, 47, 181-190.

[11] Ljunger, G., Adlercreutz, P. & Mattiasson, B. (1994). Lipase catalyzed acylation of glucose. *Biotechnology Letters*, 16, 1167-1172.

[12] Ducret, A., Giroux, A., Trani, M. & Lortie, R. (1995). Enzymatic preparation of biosurfactants from sugars or sugar alcohols and fatty acids in organic media under reduced pressure. *Biotechnology and Bioengineering*, 48, 214-221.

[13] Ku, M. A. & Hang, Y. D. (1995). Enzymatic synthesis of esters in organic medium with lipase from *Byssochlamys fulva*. *Biotechnology Letters*, 17, 1081-1084.

[14] Cao, L., Fischer, A., Bornsceuer, U. T. & Schmid, R. D. (1997). Lipase-catalyzed solid phase synthesis of sugar fatty acid esters. *Biocatalysis and Biotransformation*, 14, 269-283.

[15] Ward, O. P., Fang, J. & Li, Z. (1997). Lipase-catalyzed synthesis of a sugar ester containing arachidonic acid. *Enzyme and Microbial Technology*, 20, 52-56.

[16] Arcos, J. A., Bernabe, M. & Otero, C. (1998). Quantitative enzymatic production of 1, 6-diacyl fructofranoses. *Enzyme and Microbial Technology*, 22, 27-35.

[17] Arcos, J. A., Bernabe, M. & Otero, C. (1998). Quantitative enzymatic production of 6-*O*-acylglucose esters. *Biotechnology and Bioengineering*, 57, 505-509.

[18] Kim, J. E., Han, J. J. Yoon, J. H. & Rhee, J. S. (1998). Effect of salt hydrate pair on lipase-catalyzed regioselective monoacylation of sucrose. *Biotechnology and Bioengineering*, 57, 121-125.

[19] Adachi, S., Nagae, K. & Matsuno, R. (1999). Lipase-catalysed condensation of erythritol and medium-chain fatty acids in acetonitrile with low water content. *Journal of Molecular Catalysis B: Enzymatic*, 6, 21-27.

[20] Degn, P., Pedersen, L. H., Duus, J. Ø. & Zimmermann, W. (1999). Lipase-catalysed synthesis of glucose fatty acid esters in tert-butanol. *Biotechnology Letters*, 21, 275-280.

[21] Tsuzuki, W., Kitamura, Y., Suzuki, T. & Kobayashi, S. (1999). Synthesis of sugar fatty acid esters by modified lipase. *Biotechnology and Bioengineering*, 64, 267-271.

[22] Zhang, X. & Hayes, D. (1999). Increased rate of lipase-catalyzed saccharide-fatty acid esterification by control of reaction medium. *Journal of American Oil Chemists' Society*, 76, 1495-1500.

[23] Kelkar, D. S., Kumar, A. R. & Zinjarde, S. S. (2007). Hydrocarbon emulsification and enhanced crude oil degradation by lauroyl glucose ester. *Bioresource Technology*, 98, 1505-1508.

[24] Jia, C., Zhao, J., Feng, B., Zhang, X. & Xia, W. (2010). A simple approach for the selective enzymatic synthesis of dilauroyl maltose in organic media. *Journal of Molecular Catalysis B: Enzymatic*, 62, 265-269.

[25] Watanabe, Y., Miyawaki, Y., Adachi, S., Nakanishi, K. & Matsuno, R. (2001). Continuous production of acyl mannoses by immobilized lipase using a packed-bed reactor and their surfactant properties. *Biochemical Engineering Journal*, 8, 213-216.

[26] Watanabe, Y., Shirai, Y., Miyake, M., Kitano, J. & Adachi, S. (2013). Antimicrobial activity of monoacyl hexose coexistent with lysozyme against Gram-positive *Bacilli*. *International Journal of Food Properties*, 16, 1222-1230.

[27] Watanabe, Y., Nagai, M., Yamanaka, K., Jose, K. & Nomura, M. (2009). Synthesis of lauroyl phenolic glycoside by immobilized lipase in organic solvent and its antioxidative activity. *Biochemical Engineering Journal*, 43, 261-265.

[28] Watanabe, Y., Fukuda, T., Takahashi, N. & Adachi, S. (2014). Optimal conditions for lipase-catalyzed condensation of erythorbic acid with fatty acids in organic solvent. *Japan Journal of Food Engineering*, 15, 143-148.

[29] Watanabe, Y., Ishido, E., Fang, X., Adachi, S. & Matsuno, R. (2005). Oxidation kinetics of linoleic acid in the presence of saturated acyl L-ascorbate. *Journal of American Oil Chemists' Society*, 82, 389-392.

[30] Kobayashi, T., Adachi, S. & Matsuno, R. (1999). Synthesis of alkyl fucosides through β-glucosidase-catalyzed condensation of fucose and 1-alcohols. *Biotechnology Letters*, 21, 105-109.

[31] Suwa, N., Kubota, H., Takahashi, K. & Machida, H. (1986). Effects of sucrose fatty acid esters on spoilage of canned coffee caused by thermophilic spore forming bacteria (in Japanese). *Nippon Shokuhin Kogyo Gakkaishi*, 33, 44-51.

[32] Enda, A., Ikegami, Y., Kasetani, M. & Koike, S. (1992). Antibacterial activity of emulsifier on clostridium *thermohydrosulfuricum* (in Japanese). *Kenkyu Houkokusyo-Toyo Shokuhin Kogyo Tanki Daigaku, Toyo Shokuhin Kenkyuusyo*, 19, 167-173.

[33] Piao, J., Kawahara-Aoyama, Y., Inoue, T. & Adachi S. (2006). Bacteriostatic activities of monoacyl sugar alcohols against thermophilic sporeformers. *Bioscience Biotechnology and Biochemistry*, 70, 263-265.

[34] Onishi, T. (1992). Properties of naturally occurring antibacterial substances on the food preservation technology (in Japanese). *Seikatsu Eisei*, 36, 179-196.

[35] Nostro, A., Gemanò, M. P., D'Angelo, V., Marino, A. & Cannatelli, M. A. (2000). Extraction methods and bioautography for evaluation of medicinal plant antimicrobial activity. *Letters in Applied Microbiology*, 30, 379-384.

[36] Yanagida, T., Shimizu, N. & Kimura, T. (2005). Extraction of wax and functional compounds from fresh and dry banana leaves. *Japan Journal of Food Engineering*, 6, 79-87.

[37] Marshall, D. L. & Bullerman, L. B. (1986). Antimicrobial activity of sucrose fatty acid ester emulsifiers. *Journal of Food Science*, 51, 468-470.

[38] Sugimoto, K., Tanaka, H., Moriyama, R., Nagai, Y., Ogawa, A. & Makino, S. (1998). A loss of the antimicrobial activity of sucrose manoesters of fatty acids as caused by esterase released from the germinated spores and vegetative cells of *Bacillus cereus*. *Biocontrol Science*, 3, 17-21.

[39] Nakayama, M., Fujimoto, A., Higuchi, A., Watanabe, M., Sadakari, K., Iio, M. & Miyamoto, T. (2003). Studies on growth characteristics of *Bacillus* strains isolated from a liquid seasoning (in Japanese). *Nippon Shokuhin Kagaku Kogaku Kaishi*, 50, 537-545.

[40] Tomida, M., Suwa, N., Machida, H., Nishimura, A. & Makino, S. (1991). Inhibition of germination of *Bacillus stearothermophilus* spores by sucrose monoalkylates and other surfactants. *Nippon Shokuhin Kogyo Gakkaishi*, 38, 1044-1049.

[41] Watanabe, T., Katayama, S., Matsubara, M., Honda, Y. & Kuwahara, M. (2000). Antibacterial carbohydrate monoesters suppressing cell growth of *Streptococcus* mutans in the Presence of Sucrose. *Current Microbiology*, 41, 210-213.

[42] Moriyama, R., Sugimoto, K., Miyata, S., Katsuragi, T. & Makino, S. (1996). Antimicrobial action of sucrose esters of fatty acids on bacterial spores. *Journal of Antibacterial and Antifungal Agents*, 24, 3-8.

[43] Salton, M. R. J. (1952). Cell wall of *Micrococcus lysodeikticus* as the substrate of lysozyme. *Nature*, 170, 746-747.

[44] Hagerman, A. E., Riedel, K. M., Jones, G. A., Sovik, K. N. Ritchard, N. T., Hartzfeld, P. W. & Riechel, T. L. (1998). High molecular weight plant polyphenolics (tannins) as biological antioxidants, *Journal of Agriculture and Food Chemistry*. 46, 1887-1892.

[45] Guyot, S., Bourvellec, C. L., Marnet, N. & Drilleau, J. F. (2002). Procyanidins are the most abundant polyphenols in desert apples at maturity. *Lebensmittel-Wissenschaft und –Technologie*, 35, 289-291.

[46] Ursini, F., Rapuzzi, I., Toniolo, R., Tubaro, F. & Bontempelli, J. (2001). Characterization of antioxidant effect of procyanidins. *Methods in Enzymology*, 335, 338-350.

[47] Funayama, M., Arakawa, H., Yamamoto, R., Nishino, T., Shin, T. & Murano, S. (1995). Effects of α- and β-arbutin on activity of tyrosinases from mushroom and mouse melanoma. *Bioscience Biotechnology and Biochemistry*, 59, 143-144.

[48] Kurosu, J., Sato, T., Yoshida, K., Tsugane, T., Shimura, S., Kirimura, K., Kino, K. & Usami, S. (2002). Enzymatic synthesis of α-arbutin by α-anomer-selective glucosylation of hydroquinone using lyophilized cells of *Xanthomonas campestris* WU-9701. *Journal of Bioscience and Bioengineering*, 93, 328-330.

[49] Horowitz, R. M. & Gentili, B. (1969). Taste and structure in phenolic glycosides. *Journal of Agricultural and Food Chemistry*, 17, 696-700.

[50] Yusof, S., Ghazali, H. M. & King, G. S. (1990). Naringin content in local citrus fruits. *Food Chemistry*, 37, 113-121.

[51] Takii, H., Matsumoto, K., Kometani, T., Okada, S. & Fushiki, T. (1997). Lowering effect of phenolic glycosides on the rise in postprandial glucose in mice. *Bioscience Biotechnology and Biochemistry*, 61, 1531-1535.

[52] Hashimoto, N., Aoyama, T. & Shioiri, T. (1981). A simple efficient preparation of methyl easters with trimethylsilyldiazomethane ($TMSCHN_2$) and its application to gas chromatographic analysis of fatty acids. *Chemical and Pharmaceutical Bulletin*, 29, 1475-1478.

[53] Fujinami, Y., Tai, A. & Yamamoto, I. (2001). Radical scavenging activity against 1, 1-diphenyl-2-picrylhydrazyl of ascorbic acid 2-

glucoside (AA-2G) and 6-acyl-AA-2G. *Chemical and Pharmaceutical Bulletin*, 49, 642-644.

[54] Cunha, L. M., Oliveira, F. A. R. & Oliveira, J. C. (1998). Optimal experimental design for estimating the kinetic parameters of processes described by Weibull probability distribution function. *Journal of Food Engineering*, 37, 175-191.

[55] Fabianek, J. & Hepr, H. (1967). Antiscorbutic activity of D-araboascorbic acid. *Proceedings of the Society of Experimental Biology and Medicine*, 125, 462-465.

[56] Christen, H. R. "Carbohydrates, Principles of Organic Chemistry" O. Salle ed., 1985, p. 952.

[57] Obata, H., Tanishita, J. & Tokuyama, T. (1987). Oxidation of reductones with ascorbate oxidase (in Japanese). *Journal of the Agricultural Chemical Society of Japan*, 61, 469-469.

[58] Humeau, C., Girardin, M., Coulon, D. & Miclo, A. (1995). Synthesis of 6-*O*-palmitoyl L-ascorbic acid catalyzed by *Candida antarctica* lipase. *Biotechnology Letters*, 17, 1091-1094.

[59] Watanabe, Y., Adachi, S., Nakanishi, K. & Matsuno, R. (1999). Condensation of L-ascorbic acid and medium-chain fatty acids by immobilized lipase in acetonitrile with low water content. *Food Science and Technology Research*, 5, 188-192.

[60] Park, K. M., Lee, D. E., Sung, H., Lee, J. H. & Chang, P. S. (2011). Lipase-catalyzed synthesis of erythrobyl laurate in acetonitrile. *Food Chemistry*, 129, 59-63.

[61] Minemoto, Y., Hakamata, K., Adachi, S. & Matsuno, R. (2002). Oxidation of linoleic acid encapsulated with gum arabic or maltodextrin by spray-drying. *Journal of Microencapsulation*, 19, 181-189.

[62] Adachi, S., Ishiguro, T. & Matsuno, R. (1995). Autoxidation kinetics for fatty acids and their esters. *Journal of American Oil Chemists' Society*, 72, 547-551.

[63] Ishido, E., Minemoto, Y., Adachi, S. & Matsuno, R. (2001). Autoxidation process of linoleic acid and methyl linoleate mixed with saturated fatty acid or its methyl ester, *Lebensmittel-Wissenschaft und – Technologie*, 34, 234-238.

In: Emulsifiers
Editor: Adrienne Fitzgerald

ISBN: 978-1-63483-688-3

Chapter 2

THE IMPACT OF COMBINED EMULSIFIER ON CRYSTALLIZATION PROPERTIES OF NON TRANS FAT

Ivana Lončarević**, Biljana Pajin and Jovana Petrović
Faculty of Technology, University of Novi Sad, Novi Sad, Serbia

ABSTRACT

This research examined the influence of combination of two kind of emulsifiers and combined emulsifier 2 in 1 on crystallization properties of non trans fat.

Nuclear magnetic resonance (NMR) spectroscopy was used for measuring the solid fat content (SFC) of fat samples at different temperatures, as well as for crystallization rate under static conditions, by measuring the change of SFC as a function of time. Kinetics of crystallization was defined applying Gompertz's mathematical method. Crystallization behavior of fat was also monitored using rotational viscometry and texture analyzer, while the melting point of fat was determined by differential scanning calorimetry (DSC).

Rheological measurements and crystallization kinetics have showed that both types of emulsifiers accelerated the crystallization in relation to fat without emulsifiers. Fat samples with emulsifier 2 in 1 had a higher crystallization rate and thus formation of lower amount of smaller crystalls, compared to fat samples with combination of two emulsifiers.

---

* Corresponding author: Ivana Lončarević. E-mail: ivana.radujko@tf.uns.ac.rs.

Texture analyses also showed better spreadability of those samples in relation to samples with combination of two emulsifiers. On the other hand, the presence of emulsifiers in fat increased the melting point and melting enthalpy, especially the addition of combined emulsifier 2 in 1.

INTRODUCTION

Fat based confectionery products contain a substantial amount of fat (up to 30-40%), where a significant part is present in crystallized form. Fat contributes to the structure of the product, where many of the sensory attributes such is spreadability, mouth feel, texture, etc. are dependent on the mechanical properties of underlying fat crystal networks [1]. Fat phase completely determines the hardening and melting as well as the consistency of the final product. Therefore, the fat selection depends on its characteristics and complex processes that may occur during manufacture and later in storage [2]. The mechanical properties of edible fats depend on a series of factors, including the solid fat content (SFC), the polymorphism of the solid state, and the microstructure of the network of crystalline particles. Although SFC does not always directly affect fat hardness, the SFC profile of the oils and fats remains of considerable interest [3]. Pulsed nuclear magnetic resonance (pNMR) is the only non-destructive direct method for the determination of SFC [4]. SFC fat profile may be useful information for creating the physical properties of the product and predicting the behavior of the final product during storage and transport. NMR is the only method that directly measures the SFC as opposed to indirect methods of differential thermal analysis (DTA) and differential skaning calorimetry (DSC) where the results are obtained by measuring the heat capacity as for volume changes resulted from the melting of the solid phase [5, 6]. The kinetics of fat crystallisation, being dependent on fat composition and on the processing conditions, is important for controlling operations in the food industry in order to produce the desired product characteristics [7]. Plastic fats are normally produced by hydrogenation, fractionation, chemical transesterification or a combination of these methods which offer the possibility of changing the physical and chemical characteristics. Recently, the use of chemical transesterification is increasing within industry production for the benefit of avoiding *trans* fatty acid formation [8]. The process of modifying the plastic properties of fats by partial hydrogenation has been criticized because the resulting fats are perceived as less healthy than the native vegetable oils.

Partial hydrogenation results in the formation of *trans* fatty acids whereas most natural vegetable fats and oils contain only *cis* double bonds [9].

Beside carefully selected raw materials, industrially produced food requires emulsifiers to facilitate processing and ensure finished products with a uniform quality and long shelf-life. Food emulsifiers are surface-active lipids that display many functions in relation to food texture and rheological properties. The functional properties of emulsifiers have been studied extensively in model systems and foods [10]. Emulsifiers are functional additives used in food industry to improve texture, stability, volume, softness, aeration and shelf life. In fat-based products, they can be used, among others, to control fat crystallization properties. In particular, some emulsifiers are added as "crystal structure modifiers" and "polymorphic retardant agents," since they can crystallise together with triacylglycerols and therefore prevent or retard the polymorphic transformations. In addition, emulsifiers can also act as seeds for crystallisation by crystallizing before the triacylgycerols and thereby accelerating the nucleation rate [11]. An emulsifier reduces the interfacial tension between oil and water and therefore it facilitates the disruption of emulsion droplets during homogenization. The emulsifier adsorbs on the surfaces of emulsion droplets to form a protective coating that prevents the droplets from aggregating with each other [12]. All of emulsifiers that found their way into confectionery products have a common feature that makes them suitable as emulsifying agents; namely they are ambiphilic, possessing both lipophilic and hydrophilic properties. The nature of this property is often expressed as a Hydrophilic-Lipophilic Balance or HLB. The HLB number is an indication of the properties of an emulsifier, usually given on a scale of 0 to 20. An emulsifier with a low HLB will tend to be more oil-like and will therefore have a greater affinity for the oil phase, so it is soluble in fats and oils and insoluble in water and ethyl alcohol. Lecithin, for example, has an HLB of 4 and has an affinity for the oil phase in chocolate. By contrast, an emulsifier with high HLB value is quite soluble in water. Polysorbate 60 has an HLB of 15 and has an affinity for the syrup phase in toffees and caramels. It is often the case in confectionery products that a combination of two emulsifiers in a formulation containing two distinct phases results in a longer lasting and more uniform product. In these cases, combinations of low and high HLB emulsifiers often give the best results [13].

This research examined physical and crystallization characteristics of non trans fat with the addition of combination of two kind of emulsifiers and with the addition of combined emulsifier 2 in 1, which serves as an alternative instead of combination of two different emulsifiers.

Materials and Methods

Starting Materials

Edible fat DELIAIR™04 – intended for fat fillings, produced by Aarhus Karshamn, Sweden.

GRINDSTED® PGPR 90, polyglycerol ester of polycondensed fatty acids from castor oil, produced by Danisco, Malaysia (denoted as E_1);

GRINDSTED® CITREM LR 10 EXTRA KOSHER, citric acid ester of mono-diglyceride made from edible refined sunflower oil, produced by Danisco, Germany (denoted as E_2);

GRINDSTED® CITREM 2 IN 1 KOSHER, citric acid ester of mono-diglyceride made from edible refined sunflower oil, produced by Danisco, Germany (denoted as E_{2in1}).

Plan of Experiments

In the experimental work the various concentrations of emulsifiers were added to fat in order to get the following samples:

F0 – fat with no emulsifiers added
F1 - fat + 0.1% E_1 + 0.2% E_2
F2 - fat + 0.15% E_1 + 0.3% E_2
F3 - fat + 0.25% E_1 + 0.5% E_2
F4 - fat + 0.3% E_{2in1}
F5 - fat + 0.45% E_{2in1}
F6 - fat + 0.75% E_{2in1}

Methods

Preparation of fat samples - the mixture of fat and emulsifiers were homogenized at room temperature in Homogenizer Ultraturrax T-25 (Janke Kunkel, Germany), with a rotation speed of 6000 rpm for 5 minutes.

Fatty acid content - the fatty acid composition in fat was determined by gas chromatography (ISO 5508:1990), using gas chromatograph Becker 409 (Packard, Zurich, Switzerland), equipped with a packed steel column (3 m x 3 mm) coated with 10% SP™ 2330 stationary phase immobilized on a

Chromosorb W/AW of 60-80 mesh particle size. Nitrogen was used as an inert carrier (15 mL/min), whereas for the detection of eluted compound flame ionization detector was used. Methyl-esters were separated under isothermal regime applying the oven temperature of 170°C, while detector temperature was 250°C [14].

SFC in function of temperature - the SFC on temperatures 20, 25, 30, 35 and 40°C was determined by Bruker minispec mq20 NMR Analyzer pulse device (Bruker, Germany), according to the method ISO 8292:1991 [15].

Crystallization rate under static condition - the crystallization rate under static conditions was followed by measuring the change of SFC as a function of time by Bruker minispec mq 20 NMR Analyzer pulse device. Approximatly 3 g of melted fat were put into the glass NMR tubes and heated for 30 minutes at 50°C to destroy the crystal history. Then, the samples were placed directly in a water bath at a crystallization temperature of 20°C. SFC measurements were taken at one minute intervals within duration of one hour.

Rheological properties of crystallizing fats - rheological properties of fat samples were determined by rotational rheometer Rheo Stress 600 (Haake, Germany), using a concentric cylinder system (sensor Z20 DIN). The samples were thermostated first at 50°C and then the temperature was lowered to 10°C, monitoring the change of viscosity.

Thermal characteristics - differential scanning calorimetry (TA Instruments, US) was used to determine the thermal profile of fat samples. 5 mg of fat samples were weighed in aluminium pans and pierced covers were sealed in place. An empty, hermetically sealed aluminium pan was used as a reference. Samples were analyzed by the following procedure: they were heated from 10°C to 100°C with heating rate of 5°C per minute.

Consistency - textural properties of fat samples were analysed using a Texture Analyser TA.XT Plus (Stable Micro System, UK). The hardness and work of shearing were determined by penetration at temperature 20°C, according to analyses Margarine Spreadability-MAR4_SR.PRJ (using software Exponent by Stable Micro Systems). The accessories included TTC Spreadability Rig (HDP/SR) using 5 kg load cell and Heavy Duty Platform (HDP/90). Each sample was placed into the cone sample holder and pressed down in order to eliminate air pockets. Any excess of sample was scraped off with a knife. Then the filled cone sample holder was put in base holder and 45 degree cone probe was used to penetrate the samples. The distance between cone sample and cone probe was 23 mm with test speed of 3 mm/s.

Results and Discussion

Fatty Acid Composition

The fatty acid composition of fat is given in Table 1. DELIAIR™04 belongs to the groups of fats with no *trans* fatty acids and with small amount of lauric fatty acids (1.0%). On the other hand, it contains high level of saturated fatty acids (61.0%) and lower amount of monounsaturated (31.9%) and polyunsaturated fatty acids (7.1%) with no ω-3 fatty acids.

SFC in Function of Temperature

SFC curves of pure fat and fat with emulsifiers at the different selected temperatures are presented on Figure 1.

At 20°C the addition of combined emulsifier increased SFC of fat, where increasing the concentration of this emulsifier additionaly increased SFC in fat samples (from 36.03% in F0 to 38. 02% in F6).

On the other hand, all samples have nearly the same SFC content at temperatures 25, 30, 35 and 40°C, meaning that addition of emulsifiers do not have a big influence on this parameter.

Table 1. Fatty acid composition of fat DELIAIR™04

Fatty acid	fat DELIAIR™04
12:0	0.6
14:0	1.0
16:0	26.5
18:0	18.8
18:1	31.9
18:2	7.1
20:0	2.6
22:0	11.5
Saturated	61.0
Monounsaturated	31.9
Polyunsaturated	7.1
Trans fatty acid	n.d.

n.d. - not detected.

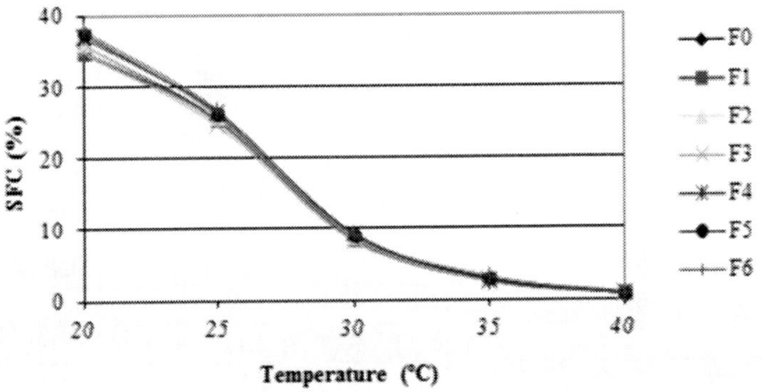

Figure 1. SFC in function of time.

Crystallization Rate under Static Condition

The crystallization properties of different fats can be compared applying Gompertz's mathematical model, which gives the dependence of the solid phase during crystallization of time under isothermal conditions [13]:

$$S(t) = a \cdot \exp\left(-\exp\left[\frac{\mu \cdot e}{a}(\lambda - t) + 1\right]\right)$$

where S is the SFC (%) at time t (min), a is the value for S when t is approaching infinity (%), μ is the maximum crystallization rate (%/min), and λ is a parameter proportional to inductive time (min). The parameters of this model were determined on the basis of experimental data by means of nonlinear regression for all fat samples. Coefficient of determination (R^2) indicate how well experimental data fit a Gompertz's mathematical model. The obtained parameters, including the estimates of the 95% confidence interval, are shown in Table 2. The data from Table 2 show that combined emulsifier 2 in 1 increases the rate of crystallization at 20°C, regardless of the amount added, while combination of two kind of emulsifiers decreases the rate of crystallization in relation to pure fat. Increasing the amount of both types of emulsifiers decreases the total amount of crystals formed, whereby fat samples with combined emulsifier 2 in 1 have lower values of parameter a at a given concentration.

Table 2. Parameters of Gompetz's mathematical model

Sample	a (%)	μ (%/min)	λ (min)	R^2
F0	26.85	4.93	1.22	0.94
F1	26.84	4.48	1.10	0.95
F2	26.74	4.59	1.04	0.93
F3	26.00	4.64	1.11	0.94
F4	26.61	6.18	2.49	0.99
F5	26.21	5.43	1.96	0.99
F6	25.58	5.73	2.12	0.99

Parameter λ indicates that inductive period is longer by adding combined emulsifier 2 in 1, while the combination of emulsifier accelerates the creation of nucleus compared to fat without emulsifiers. High values of the coefficient of determination (R^2), especially in samples with combined emulsifier indicate that the application of the Gompertz's mathematical model for describing experimental data by a theoretical curve is justified.

Rheological Properties of Crystallizing Fats

Figure 2 shows the rate of crystallization of the fat samples, depending on the type and quantity of added emulsifiers, determined by rotational rheometer. The crystallization started at temperature 35°C for all samples, followed by a rapid crystal growth in temperature range 27-23°C for fat samples with emulsifiers. The addition of both emulsifiers accelerated change in viscosity in relation to pure fat sample during the crystallization in temperature range 50-10°C, regardless to type and concentration of emulsifier.

Thermal Properties

Table 3 shows melting point of fat samples as well as energy required for crystals melting.

Although the addition of combination of emulsifiers E_1 and E_2, as well as the combined emulsifier E_{2in1} reduces the amount of crystals formed during the crystallization at 20°C in relation to pure fat, the DSC measurements showed that the presence of emulsifiers in fat increased the melting point and melting enthalpy.

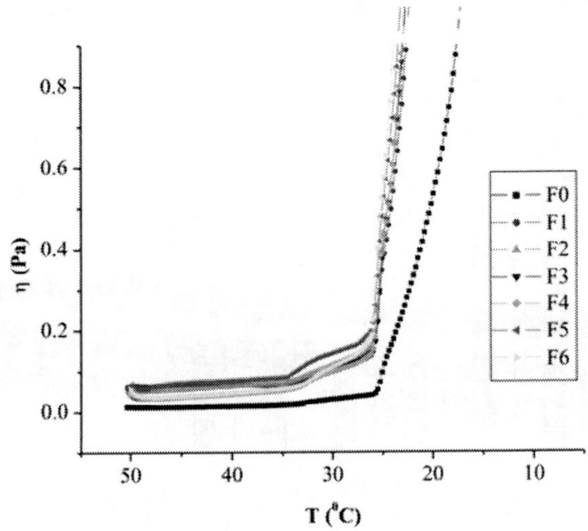

Figure 2. The change of viscosity during crystallization.

Table 3. Melting point and entalphy required for crystals melting

Sample	Melting point (°C)	Melting entalphy (J/g)
F0	32.02	23.02
F1	32.45	23.02
F2	34.74	31.70
F3	34.78	36.32
F4	34.51	29.75
F5	34.99	31.24
F6	36.11	31.37

Moreover, fat samples with combined emulsifier 2 in 1 had higher values of melting point at same concentrations, compared to samples with combination of two emulsifiers. Increasing the concentration of both emulsifiers increased melting enthalpy in comparison to fat with no emulsifiers added.

Consistency

Figure 3 represents the influence of emulsifiers on consistency of fat samples at 20°C.

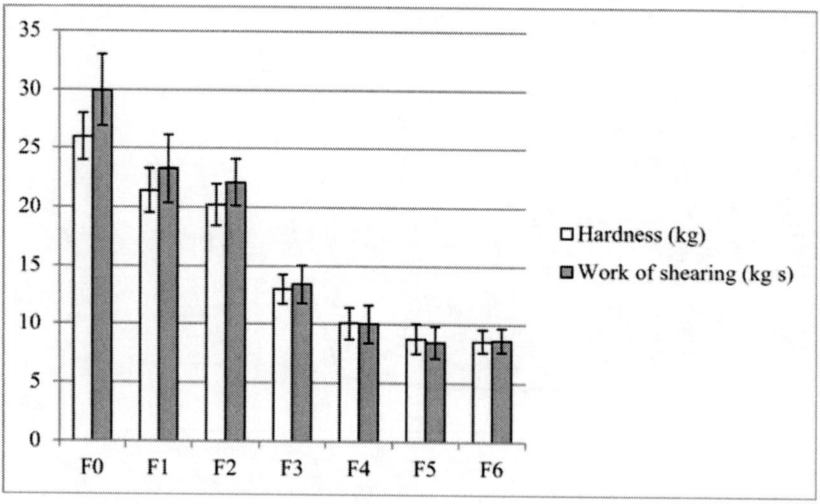

Figure 3. Textural characteristics of fat samples.

The consistency of examined fat samples with combination of E_1 and E_2, as well as the combined emulsifier E_{2in1} is in accordance with formation of solid phase during the crystallization at 20°C.

The pure fat, which formed the highest amount of solid phase during the crystallization, also has the highest hardness and work of shearing at 20°C, compared to fat samples with emulsifiers. As increasing the amount of both types of emulsifiers decreased the total amount of crystals formed, they also decreased the hardnes of fat samples where samples with combined emulsifier 2 in 1 have lower values of textural parameters at a given concentration.

CONCLUSION

The aim of this chapter was to compare the influence of combination of two kind of emulsifiers and combined emulsifier 2 in 1 on crystallization properties of non trans fat.

Both types of emulsifiers accelerated the crystallization in relation to fat without emulsifiers, as was shown by rheological measurements. Crystallization kinetics at 20°C showed that samples with emulsifier 2 in 1 had a higher crystallization rate and formation of lower amount of smaller crystalls, compared to fat samples with combination of two emulsifiers.

This was also confirmed by texture analyses which showed better spreadability of fat samples with E_{2in1} compared pure fat and samples with combination of two emulsifiers. On the other hand, the addition of emulsifier 2 in 1 increased melting point of pure fat and fat samples with the addition of two emulsifiers E_1 and E_2.

ACKNOWLEDGMENTS

This chapter has been supported by the Ministry of Science and Technological Development of the Republic of Serbia (Project no. 31014).

REFERENCES

[1] Lončarević, I., Pajin, B., Omorjan, R., Torbica, A., Zarić, D., Maksimović, J. and Švarc Gajić, J. (2013). The influence of lecithin from different sources on crystallization and physical properties of non trans fat. *Journal of Texture Studies*, 44, 450-458.

[2] Pajin, B., Karlović, Đ., Omorjan, R., Sovilj, V. and Antić, D. (2007). Influence of filling fat type on praline products with nougat filling. *European Journal of Lipid Science and Technology*, 109, 1203-1207.

[3] Braipson-Danthine, S. and Deroanne, C. (2006). Determination of Solid Fat Content (SFC) of Binary Fat Blends and Use of These Data to Predict SFC of Selected Ternary Fat Blends Containing Low-Erucic Rapeseed Oil. *Journal of the American Oil Chemists Society*, 83, 571-581.

[4] Torbica, A., Pajin, B., Omorjan, R., Lončarević, I. and Tomić, J. (2014). Physical properties of chocolate with addition of Cocoa Butter Equivalent of moderate hardness. *Journal of the American Oil Chemists Society*, 91, 39-48.

[5] Pajin, B., Radujko, I., Šereš, Z., Šoronja Simović, D., Gyura, J. and Sakač, M. (2012). Influence of low-melting milk fat fraction on crystallization and physical properties of chocolate. *British Food Journal*, 114, 868-879.

[6] Zarić, D., Pajin, B., Lončarević, I., Šereš, Z. Dokić, Lj. and Šoronja Simović, D. (2012). The impact of manufacturing process on the content

of hard triglycerides, hardness and thermal properties of milk chocolate. *Chemical Industry*, 66, 735-741.

[7] Foubert, I., Vanrolleghem, P. A., Vanhoutte, B. and Dewettinck, K. (2002). Dynamic mathematical model of the crystallization kinetics of fats. *Food Research International*, 35, 945-956.

[8] Piska, I., Zarubova, M., Loužecky, T., Karami, H. and Filip, V. (2006). Properties and crystallization of fat blends. *Journal of Food Engineering*, 77, 433-4382.

[9] Pajin, B., Šoronja Simović, D., Šereš, Z., Gyura, J., Radujko, I. and Sakač, M. (2011). Physicochemical and textural properties of puff pastry margarines. *European Journal of Lipid Science and Technology*, 113, 262-268.

[10] Radujko, I., Jurić, J., Pajin, B., Omorjan, R., Šereš, Z. and Šoronja Simović, D. (2011). The influence of combined emulsifier 2 in 1 on physical and crystallization characteristics of edible fats. *European Food Research and Technology*, 232, 899-904.

[11] Garbolino, C., Bartoccini, M. and Floter, E. (2005). The influence of emulsifiers on the crystallisation behaviour of a palm oil-based blend. *European Journal of Lipid Science and Technology*, 107, 616-626.

[12] Sakiyan, O., Sumnu, G. and Sahin, S. (2004). Influence of fat content and emulsifier type on the rheological properties of cake batter. *European Food Research and Technology*, 219, 635-638.

[13] Hasenhuettl, G. L. and Hartel, R. W. (2008). *Food Emulsifiers and Their Applications*. New York: Springer Science.

[14] ISO 5508:1990 Animal and vegetable fats and oils. *Analysis by gas chromatography of methyl esters of fatty acids*.

[15] ISO 8292:1991 Animal and vegetable fats and oils - *Determination of solid fat content - Pulsed nuclear magnetic resonance method.*

In: Emulsifiers
Editor: Adrienne Fitzgerald

ISBN: 978-1-63483-688-3

Chapter 3

FOOD-GRADE COLLOIDAL PARTICLES AS EMULSIFIERS AND STABILIZERS FOR COMPLEX COLLOIDS

*Ashok R. Patel**
Vandemoortele Centre for Lipid Science and Technology,
Lab. of Food Tech. and Engg., Faculty of Bioscience Engg.,
Ghent University, Coupure Links, Gent, Belgium

ABSTRACT

Colloidal systems such as foams, emulsions, gels and hybrid colloids such as emulsions gels and foamed emulsions are routinely encountered when dealing with food formulations. To ensure short and long term stability of these systems, a range of small molecular weight surfactants (including phospholipids, mono and di-acylglycerols etc.) and/ or polymeric stabilizers (proteins and modified polysaccharides) are commonly used as formulation aids. In both cases, the interfaces are stabilized by adsorption of molecular layers. While, small molecular weight surfactants have higher surface activity (i.e., they lower interfacial tension at low concentrations), the macromolecular stabilizers have a better adsorption efficiency (due to their large sizes, they are able to cover the interfaces efficiently, leading to the formation of stable viscoelastic interfacial layers). Recently, a third category of

* E-mail: Patel.Ashok@Ugent.be.

emulsifiers/stabilizers have generated a lot of interest where the interfaces are stabilized by formation of an adsorbed layer of solid particles in the colloidal size range. In this chapter, a brief account of such particles (including colloidal complexes, fine crystallites and inorganic particles) and their role in stabilizing complex colloid systems such as foamed emulsions, structured emulsions and bigels is discussed with the help of illustrative examples.

Keywords: Colloidal complexes, wax crystals, foamed emulsions, bigels, structured emulsions

INTRODUCTION

Many food products can be considered as carefully formulated complex colloids. For instance, ice cream which is simultaneously both an emulsion as well as a foam is a perfect example of complex colloid. It is essentially a water continuous system where coarse air bubbles (25 - 100 μm) are stabilized by aggregates of fine fat droplets (sub-micron in size) along with adsorbed micellar casein which is nanoscale in size. In addition, there are other phases including ice crystals, glassy sugars and unfrozen liquid [1]. Stabilization of such multi-phase complex systems is usually carried out by using either small molecular weight surfactants such as phospholipids, partial glycerides, polyglycerol esters, sucrose esters, polysorbates etc. or macromolecular stabilizers (proteins and modified polysaccharides) or their combinations. The affinity of emulsifier/stabilizers to the interface is defined by two parameters: a) adsorption efficiency (a measure of minimum amount of emulsifier required to saturate an interface) and b) surface activity (a measure of maximum decrease in the interfacial tension achieved when an interface is completely saturated). While, macromolecular stabilizers have a higher adsorption efficiency (due to their large size and consequently multiple binding sites), small molecular weight surfactants have a comparatively higher surface activity (due to small size they pack more efficiently, resulting in a greater decrease in interface tension).

A third category of emulsifiers which can be used for stabilization of colloids include solid particles with suitable wettability. Examples of such solid particles include colloidal complexes of macromolecule with another macromolecule (protein-polysaccharide complexes) or with a small molecular weight component (protein-polyphenol complexes), inorganic particles (colloidal silicon dioxide) and fine crystalline particles (fat crystals, wax crystals etc.) The focus of this chapter is to give a brief account of different type of complex colloids stabilized by surface active particles such as methylcellulose-tannic acid complexes, shellac-xanthan complexes, colloidal silicon dioxide and wax crystals.

Foamed Emulsions Stabilized by Methylcellulose-Tannic Acid (MCE-TA) Complexes

Complex colloids containing both oil and air as dispersed phases in water continuous medium forms basis of many regularly consumed food products such as ice-cream and whipped cream. Formulation of such systems (usually known as foamed emulsions or emulsion foams) requires stabilization of both air-water as well as oil-water interfaces along with network stabilization of the bulk phase due to the jamming of dispersed phases. For example, in commercial whipped creams, stabilization of foam is initially achieved by adsorbed emulsifier (such as proteins) and the final structure is stabilized by agglomeration of fat globules at the air-water interface. In addition, the fat globules also aggregate in the aqueous bulk phase to form a gel-like network that helps in the physical encasing of air bubbles and prevent them from coalescing [2, 3].

Generating aqueous foam in presence of a high volume fraction of 'liquid' oils (without solid fats) with high foam expansion or overrun (a measure of the volume of air incorporated in the foam) and long term stability is still a significant challenge because the oil can act as an effective anti-foaming agent [4].

Polyphenols with abundant OH groups in their structures are known to spontaneously interact with macromolecules like proteins and cellulose derivatives, resulting in the formation of ill-defined colloidal complexes with functional properties [6-9]. Colloidal complexes (in the size range of 55 - 110 nm) created from spontaneous interactions of hydrophilic cellulose derivative-methylcellulose and plant polyphenol- tannic acid (Figure 1 a-c) were found to stabilize both the air-water (foam stabilization) and oil-water (emulsion stabilization) interfaces [5].

This excellent stabilizing property of colloidal complexes was further utilized to stabilize complex colloidal dispersion (foamed emulsion) at a very high volume fraction of oil phase ($\varphi_{oil} = 0.5$) [3].

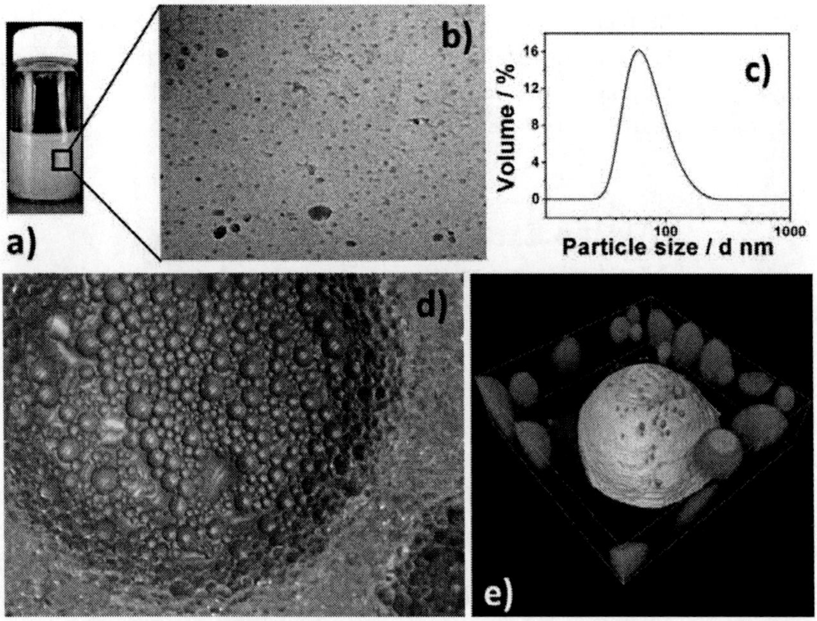

Figure 1. a) Photograph of colloidal dispersion prepared by spontaneous complexation of MCE-TA (20:1 w/w); b) TEM image of the complexes (scale width = 800nm); c) Particle size distribution of complexes; d) Microscopy image of foamed emulsion (prepared at an oil volume fraction of 0.5) stabilized by colloidal complexes and e) 3D construction from Confocal microscopy images showing that the air bubble is surrounded by fine oil droplets [3, 5].

Microscopy images of foamed emulsion (Figure 1 d and e) clearly shows that the air bubbles are stabilized by the presence of oil droplets jammed together around the air bubbles, and the coalescence of oil droplets was prevented probably by the presence of surface active colloidal complexes.

The average oil droplet and air bubble sizes in the foamed emulsion were around 20 and 200 μm respectively. The foamed emulsion was stable to structure collapse over a storage period of 4 weeks and the average droplet and bubble sizes showed only a slight increase with time. The coarsening of the foam over storage period points towards the loss of interfacial stiffness whereas, the stability against collapse suggests a close packing of the oil droplets and air bubbles in the viscous matrix of the colloidal complex. The simplicity of loading a high amount of oil in the foam phase can also be exploited to generate intensely colored foam by dissolving hydrophobic colorants such as curcumin and beta carotene in the oil phase. Such surfactant-free, richly colored foamed emulsions prepared using food-grade ingredients could have interesting applications in food formulations. This approach can further be extended to load a high concentration of hydrophobic bioactive in the foam phase for controlled delivery applications [3].

STRUCTURED EMULSIONS STABILIZED BY WAX CRYSTALS

Crystalline fat particles (high melting triacylglycerol crystals) play a very important role in providing the right texture and long term stability (to some extent) to oil continuous structured emulsions such as margarines and table spreads.

Specifically, the crystalline fat particles organize into a 3D continuous network to structure the oil continuous phase which provides the texture and the physical encasing of water droplets further prevents droplet coalescence thus, contributing to the stability [10]. The current ban on *trans* fats coupled with the increasing concerns about the negative health effects of saturated fats has meant that there is a considerable interest in formulating structured (oil continuous) emulsions in absence of solid fats (triacylglycerol crystals) [11, 12]. Wax crystals have been found to be excellent alternative to fat crystals in structuring liquid oil at a much lower mass fraction [13, 14]. Liquid oil structured using wax crystals can also be used to formulate emulsifier-free water-in-oil emulsions [15, 16]. In addition to providing texture to emulsions through bulk crystallization, the dispersed water droplets are also stabilized through interfacial crystallization (as seen in Figure 2a).

Emulsions can be formulated to mimic margarine-like consistency and melting behavior based on the concentration of wax used. A comparative time profile of emulsion at different holding temperatures is shown in Figure 2b, as seen from the value of complex modulus, G* (a measure of firmness), the emulsion can be tuned to show melting at close to body temperature [16].

SURFACTANT-FREE, PH RESPONSIVE EMULSIONS

Emulsions with on-off properties can have interesting applications in food science, particularly in the field of controlled delivery of bioactives [17].

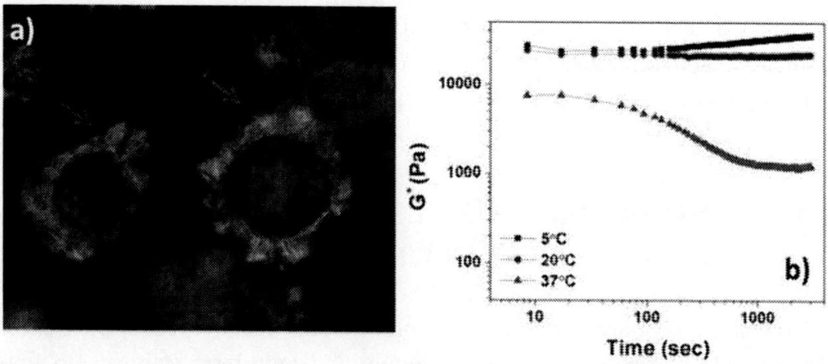

Figure 2. a) Stabilization of water droplets by wax crystals adsorbed at the interface and b) Time sweeps at 5, 20 and 37°C for emulsions prepared using 6%wt shellac wax in the oil continuous phase [16].

The possibility of controlling the properties of emulsions based on simple external factors (such as pH) could be exploited to develop novel delivery systems for oil soluble bioactives that are unstable in the acidic medium of stomach. Unfortunately, in most cases, inorganic particles are utilized for stabilization of such systems which make them unsuitable for food applications. Colloidal complexes which show pH-dependent solubility behavior were used in our lab to fabricate pH responsive emulsions. Complexes of shellac (resin) and xanthan gum (polysaccharide) were first prepared through spontaneous interactions followed by the fabrication of oil-in-water emulsions (containing up to 60% wt oil). These emulsions could be reversibly switched between stable and flocculated states by simply changing the pH (Figure 3) [18]. Based on the characterization data, the emulsions showed good stability against coalescence with no changes in the average droplet sizes and morphology, as viewed under the microscope. It can be assumed that the electrostatic repulsion of highly charged surfaces of oil droplets contributes to the enhanced stability of the emulsions.

'Aqueous-Organic' Bigels Stabilized by Hydrophilic Silica Particles

Colloidal silica particles (fumed silica) are among the most researched inorganic particles because of their complex microstructure that shows several levels of organization and interesting surface chemistry owing to the presence of surface silianol groups [19, 20].

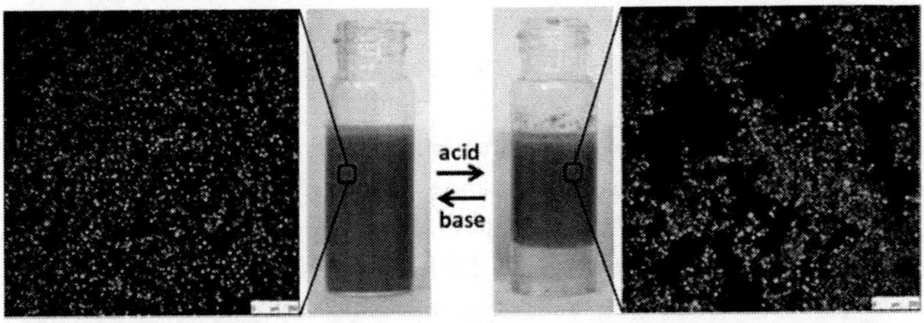

Figure 3. Photographs and confocal microscopy images (scale bars = 250μm) displaying the reversible switch between stabilized and destabilized (flocculated) emulsion by simple change of pH. Note: Oil was dyed with Nile red and xanthan gum was fluorescently labeled with Rhodamine B [18].

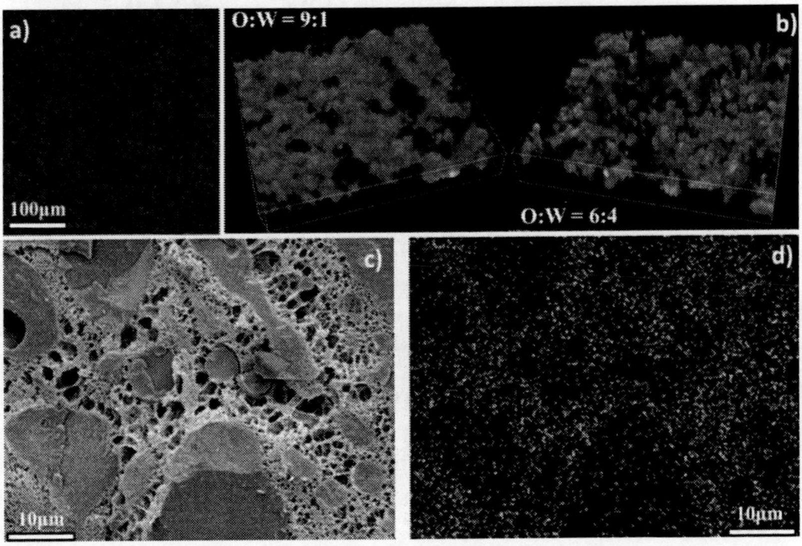

Figure 4. a) Confocal microscopy image of bigel prepared at O:W ratio of 8:2; oil phase was doped with Nile red; b) 3D volume views of bigel samples prepared at O:W ratios of 9:1 and 6:4 respectively (dimensions: x and y = 104.67 mm and z = 18 mm); (c) and (d) cryo-SEM image of bigel imaged after sublimation of water and the corresponding elemental map showing the distribution of silicon (shown in blue-green) in the bigel sample as recorded using EDS [21, 22].

Recently, hydrophilic grades of fumed silica were used as structurants to develop oleogels and a relatively unexplored class of colloids called as bigels [21]. Shear-induced de-agglomeration followed by reorganization of particle aggregates into a continuous network led to the formation of structured oil or oleogel. The structured oil system was then combined with ‘weak’ water gel (structured using hydrocolloids) under mild shear to obtain arrested de-mixed systems where both the oil and the water phases are dispersed as discontinuous networks (Figure 4a).

As seen from Figure 4 b, on increasing the proportion of water phase, the microstructure shows a distinct change from one component percolation network to a double percolation or two component percolation network. To further understand the localization of silica particles in the bigel structure, cryo-SEM imaging along with energy dispersive x-ray spectroscopy (EDS) were performed. The cryo-SEM image and the corresponding elemental map of Si from EDS are shown in Figure 4 c and d. The residual polymer mesh left after sublimation of water is clearly distinguishable from the glassy oil phase. The elemental map suggests that the silica particle preferred partitioning in the water phase with significant accumulation at the water-oil interfaces which results in stabilization of bigels.

Conclusion

To conclude, the utilization of particles (colloidal complexes, wax crystals and inorganic particles) for stabilizing different types of colloidal systems such as foamed emulsions, structured emulsions, responsive emulsions and bigels is demonstrated with the help of illustrative examples. The possibility of using food-grade components to fabricate such novel colloids could be of significant interest to the food industry for developing newer product format with advanced functionalities. However, it is important to note that the use of some of the components described here (such as fumed silica and shellac resin and wax) might face some regulatory issues as they are approved only for specific roles and not as direct additives.

References

[1] H. D. Goff, R. W. Hartel, Ice Cream Structure. Springer New York, USA, 2013.

[2] E. Dickinson, *Current Opinion in Colloid and Interface Science,* 15 (2010) 40.

[3] A. R. Patel, E. Drost, T. B. J. Blijdenstein, K. P. Velikov, *Chem. Phys. Chem.*, 13 (2012) 3777.

[4] C. A. Miller, *Current Opinion in Colloid and Interface Science*, 13 (2008) 177.

[5] A. R. Patel, J. Seijen ten-Hoorn, J. Hazekamp, T. B. J. Blijdenstein, K. P. Velikov, *Soft Matter*, 9 (2013) 1428.

[6] K. J. Siebert, N. V. Troukhanova, P. Y. Lynn, *Journal of Agricultural and Food Chemistry*, 44 (1996) 80.

[7] A. R. Patel, J. Seijen-ten-Hoorn, K. P. Velikov, *Journal of Colloid and Interface Science,* 364 (2011) 317.

[8] A. R. Patel, J. Nijsse, K. P. Velikov, *Soft Matter*, 7 (2011) 4294.

[9] C. M. Spencer, Y. Cai, R. Martin, S. H. Gaffney, P. N. Goulding, D. Magnolato, T. H. Lilley, E. Haslam, *Phytochemistry,* 27 (1988) 2397.

[10] A. G. Marangoni, N. Garti, Edible oleogels: Structure and health implications. AOCS Press, Urbana, Illinois, 2011.

[11] A. R. Patel, K. Dewettinck, *European Journal of Lipid Science and Technology,* (2015) DOI: 10.1002/ejlt.201400553.

[12] G. R. L. Dharma R. Kodali, Trans Fats Alternatives. AOCS Press, USA, 2005.

[13] A. R. Patel, M. Babaahmadi, A. Lesaffer, K. Dewettinck, *Journal of Agricultural and Food Chemistry*, 63 (2015) 4862.

[14] H.-S. Hwang, S. Kim, M. Singh, J. Winkler-Moser, S. Liu, *J. Am. Oil Chem. Soc.,* 89 (2012) 639.

[15] A. R. Patel, D. Schatteman, W. H. D. Vos, K. Dewettinck, *RSC Advances,* 3 (2013) 5324.

[16] A. R. Patel, D. Schatteman, W. H. De Vos, A. Lesaffer, K. Dewettinck, *Journal of Colloid and Interface Science*, 411 (2013) 114.

[17] R. T. Woodward, L. Chen, D. J. Adams, J. V. M. Weaver, *Journal of Materials Chemistry,* 20 (2010) 5228.

[18] A. R. Patel, E. Drost, J. Seijen ten Hoorn, K. P. Velikov, *Soft Matter*, 9 (2013) 6747.

[19] I. F. Mironyuk, V. I. Zarko, V. V. Turov, E. F. Voronin, E. M. Pakhlov, E. V. Goncharuk, R. Leboda, J. Skubiszewska-Zięba, W. Janusz, S. Chibowski, Y. N. Levchuk, A. V. Klyueva, *Journal of Colloid and Interface Science,* 242 (2001) 90.

[20] V. M. Gun'ko, I. F. Mironyuk, V. I. Zarko, E. F. Voronin, V. V. Turov, E. M. Pakhlov, E. V. Goncharuk, Y. M. Nychiporuk, N. N. Vlasova, P. P. Gorbik, O. A. Mishchuk, A. A. Chuiko, T. V. Kulik, B. B. Palyanytsya, S. V. Pakhovchishin, J. Skubiszewska-Zięba, W. Janusz, A. V. Turov, R. Leboda, *Journal of Colloid and Interface Science*, 289 (2005) 427.

[21] A. R. Patel, B. Mankoc, M. D. Bin Sintang, A. Lesaffer, K. Dewettinck, *RSC Advances*, 5 (2015) 9703.

[22] A. R. Patel, Alternative Routes to Oil Structuring. Springer International Publishing, USA, 2015.

In: Emulsifiers
Editor: Adrienne Fitzgerald
ISBN: 978-1-63483-688-3

Chapter 4

LECITHIN, MODIFIED LECITHINS, POLYGLYCEROL POLYRICINOLEATE AND SORBITAN MONOSTEARATE EFFECTS IN COCOA BUTTER AND OTHER LIPID SYSTEMS

Eriksen Koji Miyasaki, Glazieli Marangoni de Oliveira and Monise Helen Masuchi
School of Chemical Engineering,
University of Campinas, Campinas, Brazil

ABSTRACT

Lecithin, modified lecithins, polyglycerol polyricinoleate and sorbitan monostearate are emulsifiers widely used in food industry, with intrinsic abilities to modify some specific characteristic of lipid systems. In chocolate industries, standard soy lecithin and polyglycerol polyricinoleate (PGPR) are commonly added to the chocolate mass mainly for rheological adequacy. These two additives have complementary effects on viscosity and yield stress. Lecithin is reported to have greater effect on the plastic viscosity and PGPR reduces the yield stress. There are several types of lecithins and modified soy lecithins emerging as alternatives to expand the range of applications of phospholipids containing emulsifiers. The modified lecithins are produced by chemical, enzymatic or physical modification of standard lecithins. These changes affect the balance between hydrophilic and lipophilic group (HLB) of emulsifiers molecules, and hence, different effects are expected on food products. In addition, sorbitan monostearate, that is produced by esterification of sorbitol with a stearic acid, can be used as an active agent to improve the consistency and the heat resistance of oil and fat blends, currently applied for the development of low saturated lipid materials. Besides changing structural and rheological properties in food formulations, these emulsifiers also present the ability of altering lipid crystallization kinetics, and therefore, they are commonly termed as crystallization modifiers of fats. Determined by their chemical structure, emulsifiers exhibit an important role in the physical characteristics of fat-based products by delaying or speeding up the crystal nucleation, and even by changing morphology and packing density of the crystal network in the lipid matrix. Standard soy lecithins can act as crystallization promoters or inhibitors, depending on the concentration added to the fat phase. The addition of PGPR and standard soy lecithin to cocoa butter produces numerous crystals of small sizes and few crystals of larger size, respectively. Acyl–acyl interactions between additives and lipid systems are improved by similar chain length present in the fatty acids. Considering emulsifiers with structural similarities to triacylglycerol molecules, one accepted action mechanisms for explaining the fat crystals network modification is by co-crystallization. Other alternative mechanisms are considered for elucidating the variations in the crystal nucleation and growth after emulsifier additions. Sorbitan monostearate functioning in lipids is attributed to a very specific crystal network formation by the self-assemble capability of its structure. In this context, this chapter presents a comprehensive review on the structural characteristics and modification effects in lipid systems given by the addition of lecithins, polyglycerol polyricinoleate and sorbitan monostearate added to cocoa butter, palm oil and other fat blends.

1. Soybean Lecithin

1.1. General Aspects

The term lecithin is a very general and commercial designation, which describes the composition of lipids derived from components of an emulsifier of crude soybean oil. In chemical terms, the name lecithin also represents the class of phosphatidylcholines (PC) and both are considered synonymous (Raj et al. 2010).

Other description is that soybean lecithin is an emulsifier consisting of a mixture of molecules called phospholipids - lipid-based emulsifiers most commonly used in food.

Almost all fats, oils and fatty foods have this class of compounds. In some crude vegetable oils such as corn oil, rice, cotton and soybean, phospholipids content are at levels of 2 to 3% (Deman 1999).

Phospholipid may be defined as any lipid containing phosphoric acid as a mono- or diester (Belitz, Grosch Schieberle 2009). These compounds are divided in two groups: glycerophospholipids and sphingolipids (Lehninger, Nelson, Cox 2002). This chapter will focus on phospholipids from the group of glycerophospholipids, which refer to derivatives of phosphatidic acid.

Phospholipids are composed of a molecule of glycerol, fatty acids, phosphoric acid, and a second hydroxy compound, which often contains nitrogen (Belitz, Grosch Schieberle 2009, Gunstone 2001).

In the molecular structure of phospholipids, there is a hydroxyl group of phosphoric acid esterified with a residue diacylglycerol and another hydroxyl group of phosphoric acid esterified with choline, ethanolamine, serine, inositol, glycerol or diacylglycerol (Gunstone 2001). Once esterified to the phosphate group, it will result in these derivatives featuring phospholipids, being the main representatives: phosphatidylinositol (PI), phosphatidyl choline (PC), phosphatidylserine (PS), phosphatidylethanolamine (PE), phosphatidyl-glycerol (PG), and others. In the lecithin, the content of PC, PE and PI are between 29-46%, 21-34% and 13-21% respectively (Garti 2002).

Structures and abbreviations of phospholipids are shown in Table 1, based on information and schemes presented by Belitz, Grosch, Schieberle (2009), Akoh and Min (2008), Gunstone (2001) and Deman (1999).

These derivatives of phosphatidic acid are not chemical compounds with defined composition, since each of these phospholipids may contain different fatty acids in their structures, therefore, it can be considered as a group of structurally related compounds (Pokorný 2006). Lecithins are formed by mixing these derivatives of phosphatidic acid.

According to Deman (1999) and Fennema (1996), the distribution of fatty acids in phospholipids is not random, but preferably with saturated fatty acids occupying the *sn*-1 position and unsaturated fatty acids occupying the *sn*-2 position (Deman 1999).

Phosphate is often connected to the *sn*-3 position of the glycerol molecule (Akoh, Min 2008). The fatty acid composition of phospholipids is usually different from that presented in the oil that originated them.

Table 1. Structure of phospholipids

Name	Group - R_3	Chemical Structure*
General structure of phospholipids	R_3	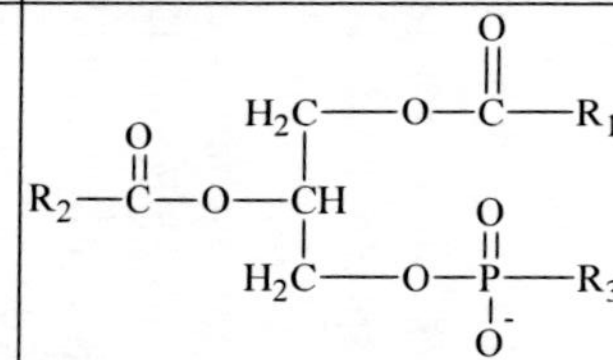
Phosphatidic acid (PA)	$-OH$	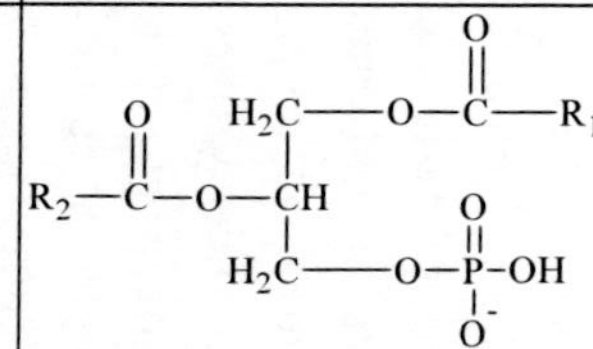
Phosphatidylcholine (PC)	$-O-(CH_2)_2N^+(CH_3)_3$	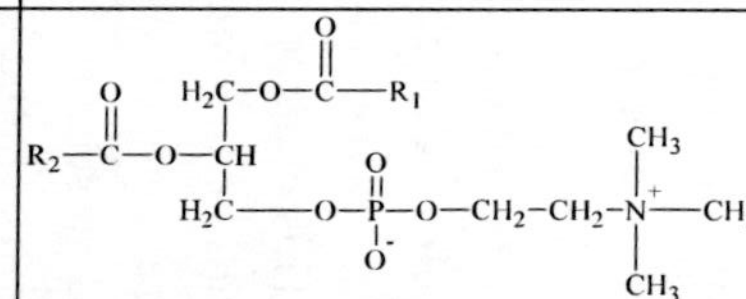
Phosphatidylethanolamine (PE)	$-O-(CH_2)_2(NH_3)^+$	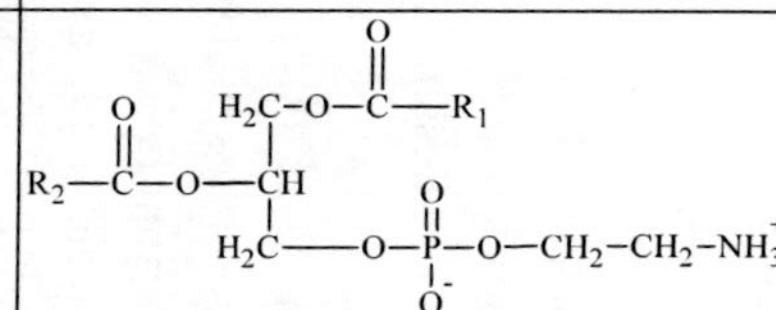

Name	Group - R_3	Chemical Structure*
Phosphatidylserine (PS)	$-O-CH_2CH(NH_3^+)COO^-$	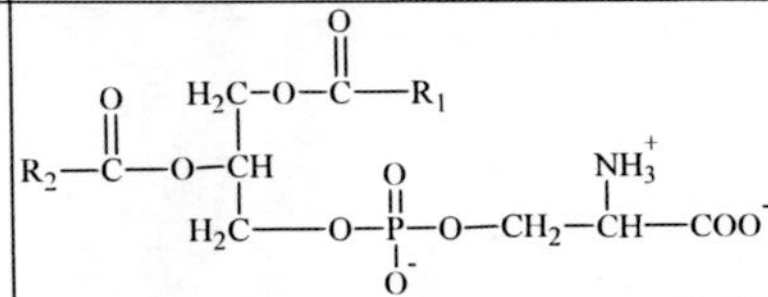
Phosphatidylinositol (PI)	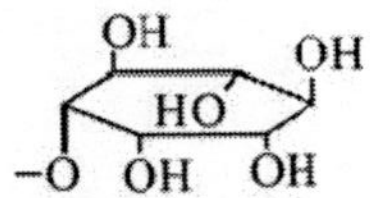 $(C_6H_{11}O_6)$	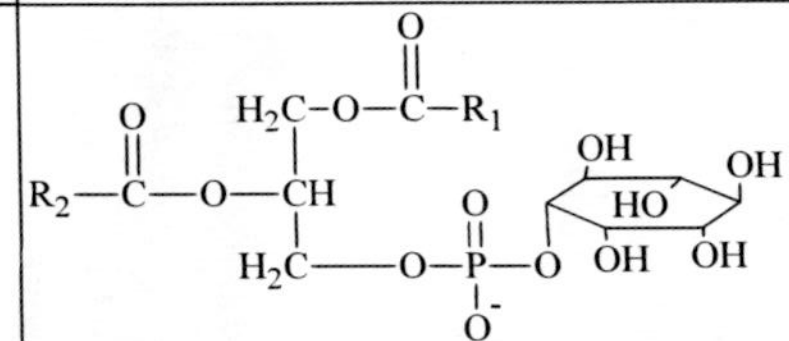
Phosphatidylglycerol (PG)	$-O-CH_2CH(OH)CH_2OH$	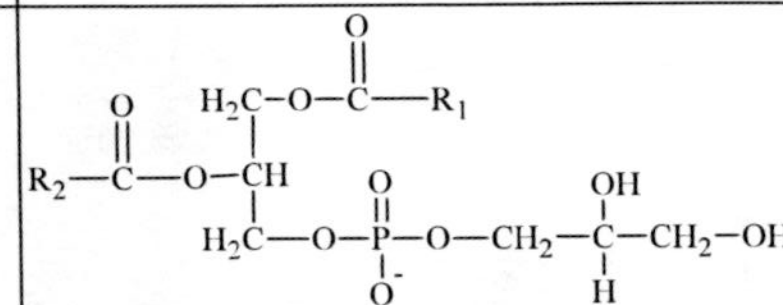
General structure of Lisophospholipids (LP)	*Phospholipid without a fatty acid*	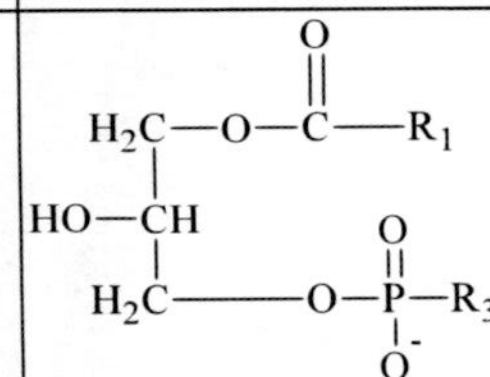

*R_1, R_2 – Fatty acids.

Table 2. Solubility lecithin and its individual components

Component	Solubility		
	Water	Alcohol	Acetone
Soybean lecithin	Insoluble	Soluble	Insoluble
PC	Soluble	Readily soluble	Slightly soluble
PE	Readily soluble	Soluble	Insoluble
PI	Readily soluble	Insoluble	Insoluble

Tanno 2012.

For having different structures in the molecule, the phospholipids have regions of different polarities with a polar zone located at the phosphate group and glycerol region, and an area related to the nonpolar fatty acid chains. These features give phospholipids the property of being surface-active compounds (Belitz, Grosch, Schieberle 2009).

The solubility of each component of soybean lecithin varies according to the type of solvent. A concentrated mixture of some of them may lead to the stabilization of an emulsion of the Water/Oil (W/O) or Oil/Water (O/W) type, even that they individually have different affinities compared to the mixture. Table 2 shows the solubility of soybean lecithin components in relation to water, alcohol and acetone.

This information can be important to elucidate the mechanisms of action of certain types of lecithin in food emulsification processes, since the molecular interactions depend on affinities in terms of polarity.

Phospholipids more easily hydratable are PC, PI and lysophosphatidylcholine (LPC), while PE and phosphatidic acid (PA) have smaller hydration capacity, and therefore, they are classified as non-hydratable phospholipids (NHP) (Nieuwenhuyzen, Tomás 2008).

The PC molecule tends to facilitate the formation of emulsions of O/W type, whereas the PE and PI (this in low concentration) tend to facilitate the formation of type W/O emulsions (McClements 2005). For type W/O emulsions, it is understood that a lecithin fraction enriched with PI is more active than one lecithin fraction enriched in PC. This is due to the type of arrangement that PI and PE molecules present at the interface (Nieuwenhuyzen, Tomás 2008, Wu, Wang 2003). For emulsions of the O/W type, different statements are known (Wu, Wang 2003).

As commercial lecithin is a mixture of phospholipids and other substances, its surface activity is the result of surface activities of all the compounds present. The range of compounds with different polarities allows that different types of lecithin not to be only particularly suitable for the type O/W or W/O emulsions, but it can operate in both cases when used in appropriate conditions of pH, temperature, and ratio of fat phase/water and formulation (McClements 2005).

For fat crystallization interpretation, understanding the mechanism of each component of lecithin is important. According to Miskandar et al. (2006), soybean lecithin can act as a promoter or inhibitor of crystallization events. Lecithins with higher proportion of polar end groups such as LPC and PI are poorly active in the nucleation and growth of crystals. On the other hand, lecithins with nonpolar end groups or lecithin concentrated in PC have been more active in nucleation.

It is observed that the literature about lecithins contains two approaches. When it comes to emulsions, lecithin rich in PI is more recommended to stabilize the type W/O emulsions (classification given to the chocolate as to type of emulsion) than the lecithin fraction with enriched with PC. However, when it comes to crystallization processes, lecithins with nonpolar end groups or PC-rich lecithins demonstrate greater influence on the crystallization (Miskandar et al. 2006). Thus, there seems to be some incompatibility to try to relate information concerning the emulsion properties, as the HLB scale or emulsion stability, with crystallization mechanisms. This fact is evidenced in subsequent topics, since the values of Hydrophilic-Lipophilic Balance (HLB) for the application of certain types of emulsifiers for emulsion (W/O or O/W) did not correlate with the crystallization behavior observed in the lipid structure. Thus, the extrapolation of information from one approach to another need to be further explained in the literature.

Studies about chocolate show that standard soybean lecithin acts as viscosity reducing agent and as inducer of crystallization. In the case of modified soybean lecithins, few studies have been found. Therefore, researches in this area could elucidate whether modified lecithins influence in the molten state or during crystallization process of the chocolate, based on the ratio of the hydrophilic and lipophilic molecules (or HLB) and/or the presence of new structures.

1.2. Production of Lecithin

The lecithin production process can be divided into two main parts: oil production and lecithin processing. In the first part during the crude oil refining process, phospholipids are almost totally (80-95%) removed from the oil during the degumming step, which usually requires a small addition of water (1-3%) and addition of an acid such as phosphoric or citric acid. This mixture is heated at 70-80°C for a certain time, promoting the precipitation of phospholipids, which are then removed by centrifugation. After this step, the phospholipids may be dried by stirring or wiped film evaporator under two conditions: 20-60 mm of film layers for 3-5 h and 60-70°C or 25-300 mm of film layers for 1-2 min and 80-105°C. These systems operate under vacuum to not cause degeneration (darkening) of lecithin due to high temperatures (Tanno 2012, Gunstone 2001).

A typical crude lecithin consisting of a mixture of 50% of phospholipids in 34% triacylglycerol, 7% of glycolipids, 7% of carbohydrates and 2% of other components. This crude product with 70-72% of acetone-insoluble content is then converted to the standard product (commercial lecithin) with 62-64% of acetone-insoluble content, acid value of 30 mg alkali/g, decreased viscosity to approximately 10^4 cP at 25°C by adding appropriate amounts of free fatty acids and triacylglycerols (Tanno 2012, Nieuwenhuyzen, Tomás 2008, Pokorný 2006, Szuhaj 2005, Gunstone 2001). However, it is emphasized that the values of these parameters may vary according to the quality standard of each manufacturer. Table 3 presents the composition of the phospholipids class derived from crude soybean oil. Table 4 presents the complete composition of soybean lecithin and, Table 5, the composition of fatty acids of soya lecithin. The crude lecithin enters the second part of the process where it will undergo modifications to increase the level of total or specific phospholipids and improve surface-active properties by structural modification. After the crude lecithin refining process, it will become the standard lecithin. This will be the raw material for producing the modified lecithin (Tanno 2012, Nieuwenhuyzen, Tomás 2008, Gunstone 2001).

Table 3. Soybean phospholipid composition - oil-free based product

Class of phospholipids	% (w/w)
Lisophospholipids (LP)	1-5
Phosphatidic acid (PA)	0-12
Phosphatidylcholine (PC)	22-46
Phosphatidylethanolamine (PE)	21-32
Phosphatidylinositol (PI)	13-22
Phosphatidylserine (PS)	5-6

Pokorný 2006.

Table 4. Standard soy lecithin composition - oil base

Components	% (w/w)
Phosphatidylcholine (PC)	19-21
Phosphatidylethanolamine (PE)	8-20
Phosphatidylinositol (PI)	20-21
Other phosphatides	5-11
Soybean oil	33-35
Sterols	2-5
Free Carbohydrates	5
Water	1

Tanno 2012.

Table 5. Composition of fatty acids of soya lecithin

Fatty acids	Carbon numbers	% (w/w)
Palmitic	C16:0	16
Stearic	C18:0	4
Oleic	C18:1	17
Linoleic	C18:2	55
Linolenic	C18:3	7
-	Others	1

Nieuwenhuyzen, Tomás 2008.

For being a co-product of crude oil processing, lecithin components may have his compositions and molecular structures altered in the process steps. Thus, changes in physical-chemical and functional properties may occur (Szuhaj 2005).

1.2.1. Production of Modified Soybean Lecithins

The modifications performed on the standard lecithin aim to improve the emulsifying properties, heat resistance and a better dispersion in aqueous or oily systems (Szuhaj 2005). The methods of modification can be divided into physical (fractionation), enzymatic and chemical (Tanno 2012, Nieuwenhuyzen 2010). Among the types of modifications, there is the fractionation of the components of crude lecithin, partial hydrolysis of the fatty acids of the phospholipid with phospholipase molecule, by partial hydroxylation of unsaturated fatty acids of phospholipid molecules and by acetylation of the amino group of phosphatidylethanolamine molecule (PE).

1.2.1.1. Fractionation

A) Fractionation by Acetone - Separation of the Soybean Oil

The neutral lipids (mainly triacylglycerols) are extracted when using acetone as solvent, and components such as phospholipids, glycolipids and carbohydrates precipitate as sediments to be insoluble in acetone (Nieuwenhuyzen 2010). After this separation, the phospholipids content in the final product is increased from 50 to 80%, the triacylglycerol decreases from 36 to 3% and the acetone-insoluble content has a value of at least 95%. As a result of this change, the viscous crude liquid turns into a powder and granular product, called powdered or defatted lecithin (Tanno 2012, Nieuwenhuyzen, Tomás 2008, Gunstone 2001).

Table 6 and 7 present the composition of phospholipids and fatty acids of defatted lecithin, respectively.

B) Fractionation by Alcohol - Separation of Enriched Fractions of PC or PI

The crude lecithin and fractioned lecithin may be defatted in terms of phospholipids with ethanol, aqueous ethanol or other alcohols. The fractionated products vary in composition and in their surfactant properties, and therefore, different applications may be possible.

Table 6. Composition of phospholipids of defatted lecithin

Components	Defatted lecithin
Phospholipids*	% (w/w)
PC	21,9
PE	13,6
PI	12,0
PS	—
PG/DPG	2,3
PA	5,8
NAPE	2,8
SPM	—
PL	2,9
Outros	3,6

Gunstone 2001.

*PL: Phospholipids; DPG: diphosphatidylglycerol; NAPE: N-acil PE; SPM: Sphingomyelin.

Table 7. Composition of fatty acids defatted lecithin

Fatty acids	Carbon Number	% (w/w)
Palmitic	C16:0	21,4
Stearic	C18:0	3,8
Oleic	C18:1	12,0
Linoleic	C18:2	57,0
Linolenic	C18:3	5,8
Arachidonic	C20:4	—
Docosahexaenoic (DHA)	C22:6	—
-	Other	—

Gunstone 2001.

Table 8. Approximate composition of some fractions of lecithin

Composition	Defatted lecithin (%)	Soluble in alcohol (%)[1]	Insoluble in alcohol (%)[2]
PC	29	60	4
PE	29	30	29
PI	32	2	55
Triacylglycerol	3	4	4
Other	7	4	8
Type of emulsion	W/O or O/W	O/W	W/O

Gunstone 2001.
[1]Enriched fraction with PC; [2]Enriched fraction with PI.

The soluble fractions in alcohol are mainly PC and PE and they act forming stable emulsions of O/W type. Insoluble fractions in alcohol are those rich in PE and PI and they act forming stable emulsions of W/O type. The proportion of PE in both enriched fractions are similar (Tanno 2012, Nieuwenhuyzen, Tomás 2008, Gunstone 2001). Table 8 presents the approximate composition of some fractions of lecithin.

C) Enzymatic Modification

The partial hydrolysis with phospholipase A_2 removes the acyl groups in the *sn*-2 position of the phospholipid and results in smooth shapes of phospholipids or simply lysophospholipids (Figure 1).

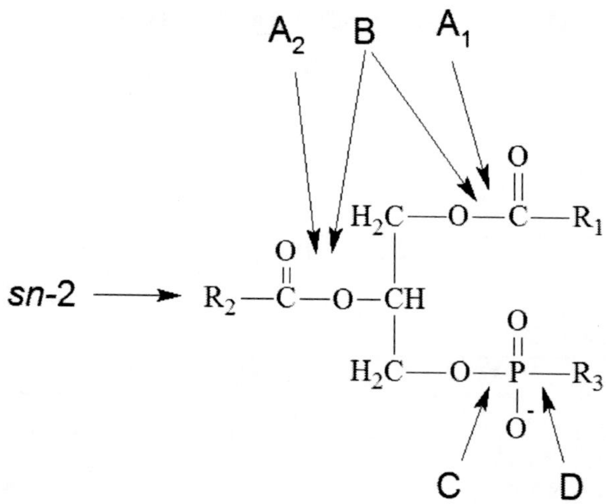

Figure 1. Phospholipid hydrolytic cleavage under catalysis phospholipases (A, B, C or D).

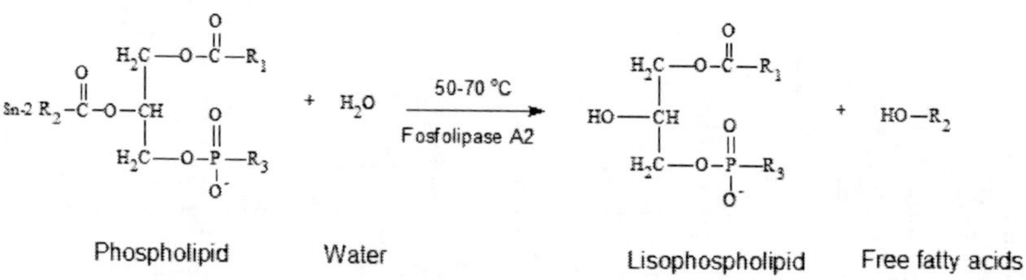

Figure 2. Enzymatic hydrolysis reaction of phospholipids.

This change resulted in products with high ratio of lysophospholipids/ phospholipids, promoting an increase of the emulsifying capacity for the emulsion stabilization of O/W type (Tanno 2012, Liu, Ma 2011, Gunstone 2001). Prior to the hydrolysis reaction, the phospholipids are subject to an adjust of moisture content, then the addition of a solution composed by phospholipase A_2 enzyme in a concentration of 0.02% (w/v) and calcium chloride in the concentration of 0.3% (w/v) is performed.

This process is carried out under stirring and at a temperature between 50-70°C for a reaction period of 7 to 9h. The hydrolysis degree reaches values of 35-40% when the acid value is in the range of 33-30 (alkali mg/g) (Liu, Ma 2011). After hydrolysis, the HLB value of 4 is altered to values between 8 to 11, relating to the standard lecithin (Belitz, Grosch Schieberle 2009). Figure 2 shows the enzymatic reaction of this modification process.

D) Chemical Modifications

Partial hydroxylation of lecithins improves the emulsion properties for systems of the type O/W. This change involves the conversion of some double bonds $(-CH = CH -)$ dihydroxide in units $(-CH\,(OH) - CH\,(OH) -)$ by reaction with hydrogen peroxide (H_2O_2) in the presence of an acid low molecular weight, such as lactic acid $(CH_3 - CH\,(OH) - COOH))$ (Tanno 2012, Liu, M. A. 2011, Xu et al. 2011, Belitz, Grosch Schieberle 2009, Gunstone 2001).

In this process, lactic acid (75%) and hydrogen peroxide (30%) are added in concentrations of 1-3% and 5-15% (v/w), respectively, to crude soya lecithin. The reaction is carried out under stirring at 50-70°C for a period of 1-3h. Subsequently, the product is neutralized with sodium hydroxide and dried under reduced pressure (Liu, Ma 2011).

Figure 3 shows the hydroxylation reaction of a fatty acid at the *sn*-2 position, preferential position of the unsaturated fatty acids. This reaction scheme shown below was built based on isolated information presented in the literature (Tanno 2012, Liu, Ma 2011; Xu et al. 2011, Belitz, Grosch, Schieberle 2009, Gunstone 2001).

In this hydroxylation process, the PE of ethanolamine group is also modified. Due to these changes, lecithin acquires appropriate properties to emulsions of the O/W type, with increasing HLB values to a range between 9-10, iodine value (IV) decreasing to 10-25 g I_2/100 g fat and increasing of hydroxyl number (Liu, Ma 2011).

For the acetylated lecithin, an acetic solution of 1-4% (v/w) $(CH_3 - CO - O - CO - CH_3)$ is added to crude lecithin under stirring and heating between 60-70°C, with a reaction period of approximately 1-1.5 h (Liu, Ma, 2011). In reaction, the free NH_2 groups present in PE molecules are acetylated (NH-Ac), becoming N-acilfosfatidiletanolamina (Tanno 2012, Xu et al. 2011, Liu, Ma 2011, Gunstone 2001). This amino group of PE, when acetylated, receives an acetyl group in the part of molecule with positive charge, which converts it to a negatively charged lecithin (SZUHAJ 2005). Thus, the zwitterionic structure is modified for obtaining an increase of Hydrophilic-Lipophilic Balance (HLB) values, improvements in thermal stability, emulsification properties to the systems of the O/W type and the viscosity properties (Liu, Ma 2011, Gunstone 2001).

After acetylation, the mixture is neutralized with sodium or potassium hydroxide and dry at vacuum. This type of lecithin has characteristics HLB values of 5-6, pH of 6.5-8.0, and 0.7 to 1.7% free amine (Xu, Wang, Liu 2008).

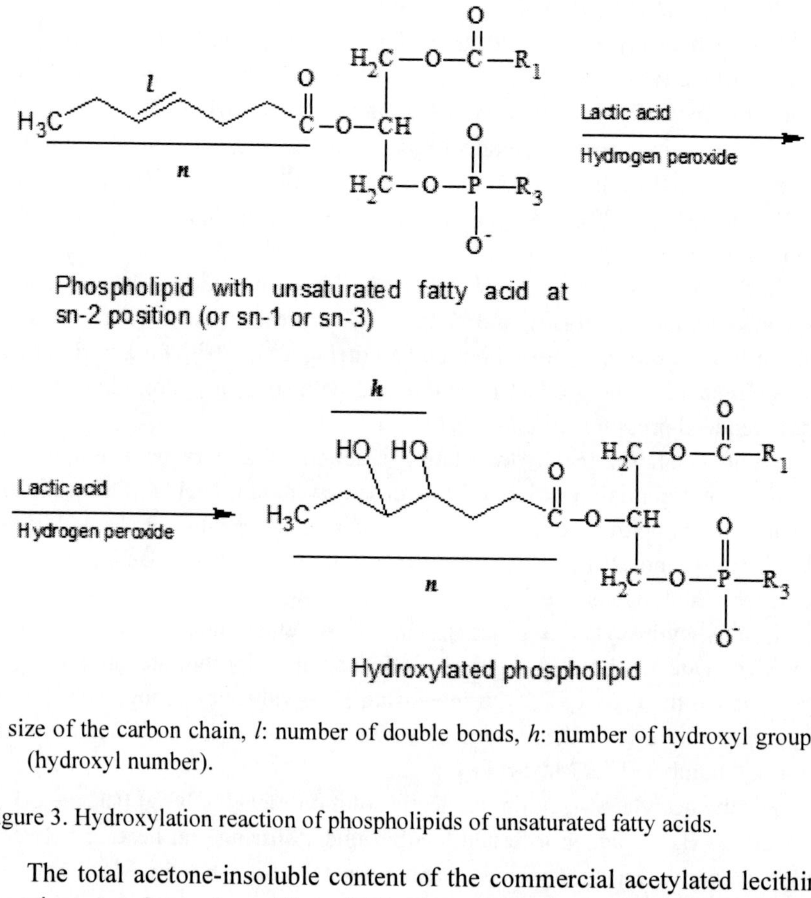

n: size of the carbon chain, *l*: number of double bonds, *h*: number of hydroxyl groups (hydroxyl number).

Figure 3. Hydroxylation reaction of phospholipids of unsaturated fatty acids.

The total acetone-insoluble content of the commercial acetylated lecithin may increase from about 52 to 97%, the remainder consisting of soy oil, natural pigments, sterols and other minor constituents present in the crude soybean lecithin (Szuhaj 2005).

The acetylation reaction is shown in Figure 4, based on information presented in the literature (Tanno 2012, Liu, M. A. 2011, Xu et al. 2011, Belitz, Grosch, Schieberle 2009, Nasir, Bernards, Charpentier 2007, Gunstone 2001).

Table 9 shows the HLB value of lecithin modified according to the literature (Nieuwenhuyzen 2010). It is emphasized that these values vary depending on the manufacturer and methods of determination of HLB.

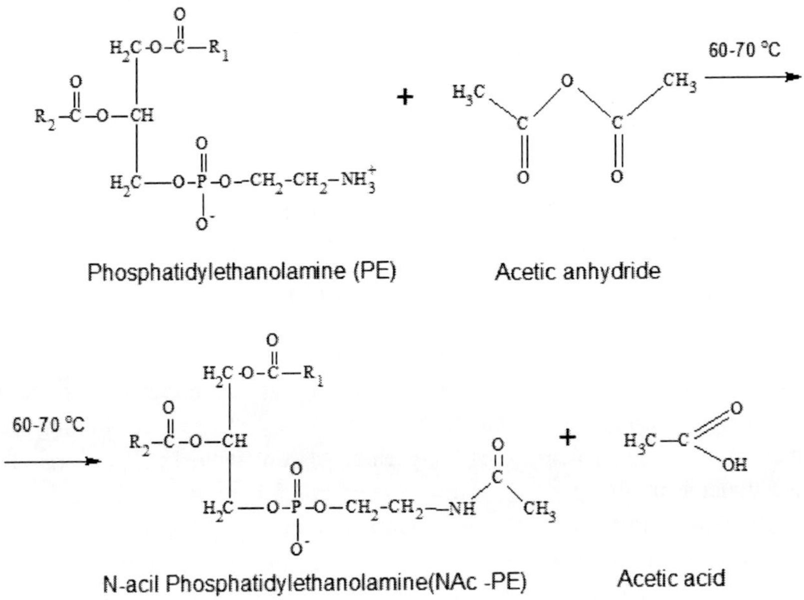

Figure 4. Acetylation reaction of phosphatidylethanolamine (PE).

2. Polyglycerol Polyricinoleate (PGPR)

Polyglycerol polyricinoleate (PGPR) is a surface active agent very different from the others. This compound was originally developed for use in the baking industry and can be obtained by polycondensation of castor oil and glycerol by reacting fatty acids with glycerol polymers containing from 2 to 10 molecules (Beckett 2000).

PGPR is one of the most hydrophobic emulsifiers used in foods, with values of Hydrophilic-Lipophilic Balance (HLB) of about 1.5 (Tisoncik 2010). This additive has gained attention recently because of its use approval in fats for confectionery, especially in products made from chocolates (Bastida-Rodríguez 2013). Its chemical structure is shown in Figure 5.

The functionality of PGPR in chocolate has been mainly associated with reduced consistency (yield value) and the ability to reduce or even cancel the yield stress. Its synergistic effect with the lecithin has lately been documented, but the specific mechanisms of action have not been fully clarified.

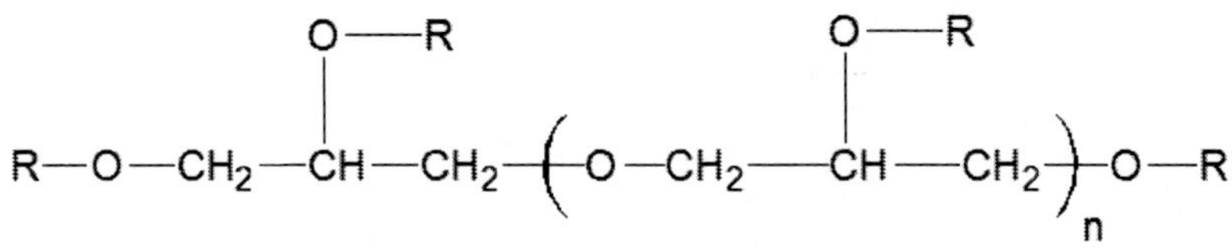

R = H or an acyl group derived from ricinoleic acid polycondensate; n = degree of glycerol polymerization (Belitz, Grosch Schieberle 2009).

Figure 5. Chemical structure of polyglycerol polyricinoleate (PGPR).

Schantz and Rohm (2005) pointed out that the use of PGPR could indirectly present significant influence in the fat crystallization, since it reduces significantly the viscosity of the chocolate mass, facilitating the achievement of better degrees of tempering. Also, it requires less energy for mass transfer in the process, allowing obtaining a product with a higher melting point, and therefore, a possible prevention of fat bloom - a common defect in chocolate products, characterized by appearance of whitish spots and film in the surface and brightness loss due to local lipid recrystallization. However, the PGPR molecule as a crystallization controller in fatty systems is pretty much still unknown and, therefore, more fundamental studies to explain their influence are required (Lonchampt, Hartel 2004, Garti 2002). In later topics of this chapter, some studies about the influence of PGPR on the crystallization of fats and rheological properties will be considered.

3. Sorbitan Monoesters

Applied in various food products, emulsifiers can be partial esters of fatty acids of animal or vegetable origin and polyvalent alcohols such as glycerol, propylene glycol, sorbitol, and sucrose. They can also be esterified with organic acids such as tartaric, lactic, succinic, and citric acids. Lecithin is the most commonly emulsifier used for chocolate production, typically in combination with PGPR - polyglycerol polyricinoleate. Both act in synergism changing the plastic viscosity and the yield stress in the chocolate rheology. However, other emulsifiers such as sorbitan esters can also be applied in the manufacturing of chocolate. They are considered less effective in reducing the yield stress or the plastic viscosity compared to mixtures of lecithin and PGPR, but they modify characteristics such as brightness, improve palatability and particularly, the shelf life of chocolates, avoiding the formation of fat bloom (Beckett 2008). These compounds promote modifications in the surface properties of lipids, resulting in changes on the size and morphology of the crystals and crystal density (Garti 2002).

Sorbitan esters are obtained by esterification of fatty acids with sorbitol-dehydrated molecule. In the case of sorbitan monoesters, there is just only one esterification of a single fatty acid at the sorbitol molecule. For example, sorbitan monoesters can be sorbitan monolaurate, monopalmitate, monostearate, and monooleate, depending on their esterified fatty acid. The general chemical structure of a sorbitan monoester is presented in the Figure 6, on which the letter R represents the carbon chain derived from different fatty acids as stearic, oleic, palmitic, and lauric acids. Sorbitan esters - mono- di- and triesters - are classified as hydrophobic emulsifiers together with mono- and diacylglycerols of fatty acids, and PGPR (Garti 2002).

In general, sorbitan monoesters exhibit effective stabilization in the polymorphic β' form in margarines and solid fat content modifications of lipid blends leading to melting profiles at the body temperature (O'Brien 2004).

In order to evaluate the emulsifier effects in fat crystallization, two different mechanisms have been reported in the literature. The first refers to the action of these additives as hetero-nuclei, accelerating the crystallization by direct catalytic action as impurities. During crystal growth, emulsifiers would be adsorbed at the crystals surface and, hence, would modify the rate of incorporation of triacylglycerols and crystal morphology (Ribeiro et al. 2015, Garti 2002, Cerdeira et al. 2003).

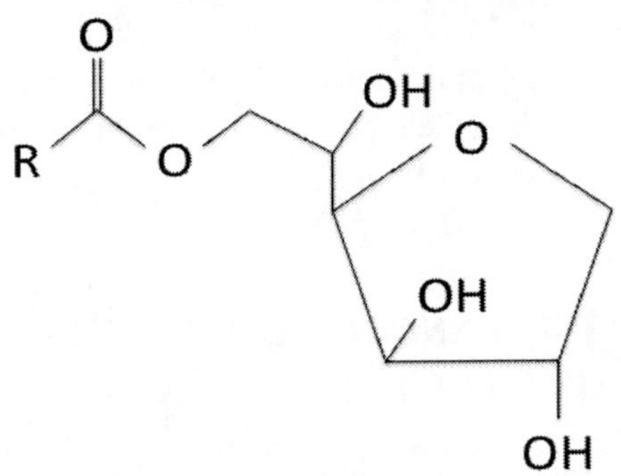

The letter R indicates the carbon chain possibilities originated from fatty acids substitutes as stearic acid for sorbitan monostearate, palmitic acid for sorbitan monopalmitate, oleic acid for sorbitan monooleate, and lauric acid for sorbitan monolaurate.

Figure 6. General chemical structure of sorbitan monoester.

The second mechanism that is of greater consensus among various authors believes that the triacylglycerols and emulsifiers would be subject to co-crystallize due to the similarity between their chemical structures (Lonchampt, Hartel 2004). Thus, a possible structural dissimilarity between triacylglycerol and emulsifier molecules can also promote a delaying in the nucleation and crystal growth, inhibiting the crystallization formation (Cerdeira et al. 2003, Garti 2002). Besides these two mechanisms extensively spread in the literature, there is a third mechanism also being considered for explaining other emulsifier mechanism in lipid matrix. Specific studies of sorbitan monostearate effects demonstrated that this emulsifier is able to form gel in organic solvents and in vegetable oils, after being dispersed and/or dissolved at high temperatures (~ 60°C). Considered as a surfactant, as the temperature of the system is cooled down, this emulsifier destabilizes the solution in which has been dissolved due to the difference in the polar affinity with the lipid phase. Promoted by this destabilization mechanism, sorbitan monostearate self-assembles in associated tubular vesicles, forming a very specific three-dimensional network (Murdan et al. 1999, Co, Marangoni 2012).

Sorbitan monostearate (SMS) has been often applied together with polysorbates in baked food products as cakes, cakes mixtures, whipped toppings, fillings, and also in confectionary coatings (Garti 2002). Besides these applications, there are some recent studies that apply SMS as a structuring modifier agent of lipid systems - as a potential tool for developing low saturated fatty acid blends - and also as a anti-blooming agent, for delaying the fat bloom formation in chocolates (Oliveira et al. 2015 (b), Yoshikawa 2014, Peyronel, Marangoni 2014, Masuchi et al. 2014, Lonchampt, Hartel 2004). Besides these applications in food technology, SMS is also being applied to aid building drug delivery systems for pharmaceutical products (Shah et al. 2013, Poletto et al. 2015).

4. INFLUENCE OF EMULSIFIERS ON FAT CRYSTALLIZATION

In addition to their known functions as emulsification and stabilization of emulsions, emulsifiers can also modify the behavior of the continuous phase of a food product, providing it with specific benefits. In lipid-based products, emulsifiers may be used to control or modify the crystallization properties of the fat phase.

In this context, the crystallization of lipids can be affected by the addition of emulsifiers with different hydrophobic properties and they may act by accelerating or slowing down the crystallization process. Additionally, emulsifiers are said to influence the polymorphic transition, size, morphology and density of the lipid crystals, and inhibit or delay the appearance of fat bloom (Hasenhuettl 2008, Garti 2002).

The study of emulsifier effects in lipid systems is of great interest to improve industrial fat-based products, particularly with respect to fat development for use in chocolate, confectionery and baking products. However, their role as crystallization modifiers in fats has been rarely investigated in the literature (Hasenhuettl 2008).

In general, the emulsifier effects are related to different crystalline organizations that can slow down the polymorphic transformations through steric hindrance or, in other cases, they can promote these transformations by favoring molecular displacements (Aronhime et al. 1987). In the literature, as mentioned before, two different mechanisms have been often reported in order to elucidate the influence of emulsifiers on the crystallization of fats.

Some examples of emulsifiers with higher potential for crystallization control of lipid blends are sorbitan esters, fatty esters and polyesters of sucrose, chemically modified lecithin and commercial lecithin pattern, and polyglycerol polyricinoleate (Lonchampt, Hartel 2004). The literature shows that these emulsifiers can affect the induction time of crystallization, nucleation and crystal growth rates and the polymorphic transition of the crystals (Weyland, Hartel 2008).

Lecithins and Polyglycerol Polyricinoleate (PGPR)

There are few studies that used PGPR as crystallization inducer, but most of them, the soybean lecithin. Bowser (2006) assessed by polarized light microscopy the influence of the addition of soy lecithin and PGPR to the cocoa butter crystallization. The results showed that lower induction times were obtained when both emulsifiers were added.

The results also showed that pure cocoa butter crystals displayed larger spherulites, with approximately the mean size of 968 μm, while the sample containing lecithin showed smaller spherulites crystals, with approximately the size of 460 nm. The sample containing PGPR was the one with the smallest size. Complementary tests performing tempering of the samples showed that the samples containing emulsifiers had a higher density than the crystalline pure cocoa butter. Rousseau et al. (2005) also achieved the same result when crystallizing hydrogenated fat cotton with PGPR.

Killian and Coupland (2012) evaluated the effect of soy lecithin and PGPR on emulsion stabilizing composed of water and soybean oil. Through optical micrograph it was found that the emulsions containing PGPR presented droplet sizes much lower than those obtained by lecithin addition, with values of approximately 21.4 m and from 43.1 to 77.0 µm, respectively. Similar results of PGPR effects in emulsions are also found in Ushikubo and Cunha (2014).

These results for the effect of PGPR on emulsions can be complementary to those of Bowser (2006), exhibited by polarized light microscopy, wherein the PGPR is stated to induce the formation of many small crystals with high density.

When the PGPR is used alone in fat, it has been claimed to have very similar crystallization behavior to pure fat. When compared with the sample of soy lecithin, PGPR induces lower crystallization rate (Wang et al. 2011).

The PGPR is stated to limit fat bloom (Bastida-Rodríguez 2013), but also control the polymorphic transition, mainly due to their different structural feature, which when set in the crystal lattice, does not change the short and long spacing values compared to other emulsifiers (Rousseau et al. 2005, Aronhime, Sarig, Garti 1987). Complementing, the PGPR molecule is also stated to act stabilizing the system around sterically (Afoakwa 2010). Based on this information and theory presented by Aronhime, Sarig, Garti (1987) on the control of molecular mobility by emulsifiers, the PGPR could act controlling the polymorphic transition and possibly the fat bloom formation.

Some studies show meticulously the effect of soy lecithin on the different ingredients of a model chocolate. Svanberg et al. (2011) analyzed the effect of lecithin on crystallization from samples containing cocoa butter, with or without particulate sugar, cocoa solids and soybean lecithin. To measure the increase in percentage of crystals over time, they used the technique of confocal laser scanning microscopy (CLSM). These authors proved that soybean lecithin also has an effect on the microstructure of samples containing cocoa solids as the sole solid component or in combination with sugar. When lecithin was present, always a higher crystallization rate was verified.

Svanberg et al. (2011) suggest a possible explanation for the increase in crystal growth when cocoa solids were present, stating that the surface of these solids has a more hydrophobic character due to the presence of fat in its structure. It could mean that the phospholipids would act as articulating agent for chocolate components, combining the polar part (phosphate group and the glycerol backbone) with the sugar particles and the polar part (fatty acids) with the fat of cocoa butter and cocoa liquor.

The unique study using modified soybean lecithin on cocoa butter crystallization is presented in Miyasaki et al. (2012). The authors studied the effect of modified soybean lecithin (acetylated, hydroxylated, enzymatically hydrolyzed and defatted) and PGPR on cocoa butter crystallization in different concentrations (0.2, 0.5 and 0.8% w/w). The outcomes were compared with the data obtained using standard lecithin and PGPR. The crystallization behavior under constant temperature were determined by Nuclear Magnetic Resonance (NMR) at the temperature of 15°C, with the sample previously melted at 60°C during 1h. From the data, Avrami parameters were determined to better elucidate the influence of these emulsifiers on the cocoa butter crystallization. The addition of modified lecithins in different concentrations changed the induction time and the Avrami parameters. Samples with lower concentration (0.2% w/w) showed an effect more pronounced or similar to those with a concentration of 0.5% in relation to the crystallization rate. Samples with highest concentration of emulsifiers exhibited curves that approached those of cocoa butter without any emulsifier. Different concentration of PGPR did not affect cocoa butter crystallization process and all samples containing this emulsifier had similar cocoa butter crystallization behavior. Samples with 0.2% of emulsifier enabled a better differentiation between effects of each one. In this concentration, among the emulsifiers verified, the enzymatically hydrolyzed lecithin was the most effective in accelerating the crystallization (according to Avrami parameter k and also observed graphically), followed by hydroxylated, standard, defatted and acetylated lecithin, and PGPR. It is important to point out that modified soybean lecithin has higher values of Hydrophilic-Lipophilic Balance (HLB) than standard soybean lecithin as previously mentioned. Literature states that emulsifiers with higher HLB values are more suitable to be applied to O/W emulsions (Hasenhuettl 2008, Miskandar et al. 2006 and Garti 2002). However, the addition of modified lecithin displayed effects at least similar to traditional standard lecithin.

These results indicated that HLB scale cannot be correlated directly with crystallization mechanisms, i.e., standard lecithin has a HLB value considered more appropriate for oily products (HLB = 4) but it has less performance than hydroxylated and enzymatically hydrolyzed lecithins (HLB = 9-10 and HLB = 8, respectively). The outcomes indicated that the stronger the polar fraction in the emulsifier molecules, the greater might be its effect on cocoa butter crystallization.

Sorbitan Monostearate

There are few studies in the literature referred to the effects of sorbitan monostearate addition on lipid systems. Some of these researches were reported by the following authors: Oliveira et al. (2015, b), Yoshikawa (2014), Peyronel and Marangoni (2014), Masuchi et al. (2014), Masuchi et al. (2012), Lonchampt and Hartel (2004), Shah et al. (2013) and Poletto et al. (2015).

Masuchi et al. (2012) evaluated the effects of sorbitan monosters - monolaurate, monopalmitate, monostearate and monooleate - addition in the crystallization behavior of cocoa butter. Sorbitan monostearate (SMS) pointed out as being the most effective structuring agent among all emulsifiers assessed - a performance assigned to its solubility in organic medium and ability of self-assembling. Adding 0.5% of SMS to cocoa butter promoted a sharp increase in the onset of the crystallization temperature (from 19.3 to 21.6°C, verified by differential scanning calorimetry) and a 60% increase in the consistency measured at 10°C. Even the classic two-step isothermal crystallization behavior noticed for cocoa butter isothermal crystallization at 17.5°C by NMR was smoothed out, showing that a polymorphic transition to a more stable crystal structure was accelerated by SMS addition.

In a later study, Masuchi et al. (2014) presented a comparison study for the effects of sorbitan monostearate and sorbitan monooleate as crystallization and consistency modifiers in cocoa butter. As the main triacylglycerols of cocoa butter are formed by palmitic, oleic and stearic fatty acids, emulsifiers containing these fatty acids such as sorbitan monostearate and monooleate, could be considered as crystallization modifiers of cocoa butter by co-crystallization of their similar molecules. However, in this study, the emulsifiers sorbitan monooleate and sorbitan monostearate exhibited marked differences in their structuring ability and, therefore, the mechanism that appeared to explain the events reported was related to other causes than the previous two mechanism presented in the beginning of this section.

The authors presented an alternative mechanism for explaining the crystallization behavior modification observed in cocoa butter by the SMS addition. This explanation was first exhibited by Murdan et al. (1999), when these authors described that the sorbitan monostearate was capable of forming a gel when dissolved or dispersed in certain organic solvents such as hexadecane, cis-decalin, trans-decalin, isopropyl myristate and also in some vegetable oils. They added 10% (w/v) of SMS in corn oil and this mixture was characterized as an opaque solution at 60°C, suggesting a weak attraction between SMS and the organic phase. The gel structure was formed when the temperature was decreased and led to a reduction of the SMS solubility in the solution. The SMS gel formed was evaluated by optical microscopy, differential scanning calorimetry, and X-ray diffraction. Sorbitan monostearate became less soluble in the organic solution when the solution temperature was decreased. Then, SMS molecules organized themselves into tubular shaped rod after increasing repulsive forces in the system, leading to a three-dimensional network formation with trapped solvent inside its structure (Murdan et al. 1999). According to the authors, the SMS network acts like a frame that preserves and keeps the lipid system inside its structure and a possible amorphous fat phase can be maintained. Masuchi et al. (2014) also observed the formation of an opaque solution at 60°C, suggesting that the cocoa butter and SMS also exhibit some degree of insolubility as reported by Murdan et al. (1999), favoring the self-assembling characteristics previously assigned to SMS and forming a very unique three dimensional structure. Consequently, it was proposed that the addition of SMS significantly changes the microstructure of the cocoa butter to a more dense and temperature resistant crystalline phase, explained by this peculiar mechanism as evidenced by the results reported. This phenomenon can be applied to develop fat products with higher resistance to temperature increases and oscillation.

Sorbitan Monostearate in Low-Sat Fat Systems

Vegetable oils and fats, as found in nature, feature some restrictions for direct applications in processed foods. An alternative is adequate the technological properties of the formulations of raw materials using advanced technology. Chemical interesterification and thermal fractionation processes are widely used in food industry for modifying the functional attributes of oils and fats, besides being an alternative to "zero trans" product formulations. However, these products require higher production costs.

Other technique to modify a lipid system contemplates the addition of structuring agents, as emulsifiers. These additives are used in development of fat-based products, composed by low content of saturated fatty acids (*low sat*), nevertheless are able to maintain functionalities equivalent of commercial partially hydrogenated fats. *Low sat* fats are associated to nutritionally adequate food, in view of the fact, interact positively with health issues.

Some emulsifiers added to oils and fats can behave as modifiers in the crystallization process and are able to stabilize specific polymorphic habits (Garbolino, Bartoccini, Flöter 2005). Moreover, lipid components containing fatty acids similar to the acyl groups in the fat change the structural properties (Smith et al. 2011, Oliveira et al. 2015, a). Different categories of sorbitan monoesters provide a variability in the chemical compositions, including sorbitan monolaurate, sorbitan monopalmitate and, sorbitan monostearate.

Sorbitan monostearate is able to arrange systems known as oleogels, which represents liquid oils entrapped in a network established by solid lipid materials formed through a self-assembly mechanism. Moreover, the sorbitan monostearate can act as crystals particles in the nucleation stage and with subsequent crystals growth inside the oil phase (Dassanayake, Kodali, Ueno 2011).

Oliveira et al. (2015, b) studied blends of palm oil and canola oil, as *low sat* lipid systems. The functional attributes of these blends were adjusted by the incorporation of sorbitan monostearate and fully hydrogenated canola oil. The authors reported a modification in crystallization profile and effectiveness in the structuration of unsaturated triacylglycerols, resulting an oleogel. The incorporation of high melting point triacylglycerols to sorbitan monostearate can favor the formation of a homogeneous crystal network, by structuring dispersed crystals, providing plasticity and spreadability to the lipid system. Furthermore, sorbitan monostearate proved to provide favorable characteristics for application in lipid-based products.

5. Influence of Emulsifiers on the Rheological and Texture Properties of Chocolate

Emulsifiers are active agents at interfaces, usually added in originally immiscible systems, which have a lipophilic phase and another hydrophilic phase. In the case of chocolate, the emulsifier acts on the dispersion of the solid phase (sugar, cocoa solids) in the continuous lipid phase (cocoa butter), inhibiting the agglomeration of fat and decreasing the viscosity (Beckett 2008, Lonchampt, Hartel 2004).

The rheological behavior of chocolate formulation is variable and dependent on several factors. Afoakwa, Paterson, Fowler (2007) stated that the rheology features of the chocolate depends on processing, ingredients of the solid particle size, type and concentration of emulsifiers, degree of saturation of the fat phase, among other factors. The rheological profile determines the efficiency of the mixing, pumping and transport of the chocolate mass during processing.

Many mathematical models may describe the flow behavior of chocolate, however, the most widely used model is the Casson model (Equation 1). The obtained parameters from this model are the yield stress and the plastic viscosity.

$$\sqrt{\sigma} = \sqrt{\sigma_{CA}} + \sqrt{\eta_{CA}} \cdot \sqrt{\dot{\gamma}} \quad (1)$$

where σ is the shear stress (Pa), σ_{CA} is the *yield stress* (Pa), η_{CA} is the Casson plastic viscosity (Pa.s), $\dot{\gamma}$ is the shear rate (s^{-1}).

According to Quiñones-Muñoza et al. (2011), the yield stress is the force required to initiate the flow. The plastic viscosity is a parameter which describes the ability to keep the fluid in movement, thereby determining pumping characteristics, coating properties and sensory characteristics of the final product, such as flavor and presence of air bubbles. The use of the Casson model to calculate rheological parameters requires tests to be carried out with shear rates (shear rate) between 2-50 s^{-1}. The shear stress values (shear stress) at shear rates (shear rate) 5^{-1} s represent the yield stress values. The plastic viscosity and apparent viscosity are determined at shear rate of 30 s^{-1} (Afoakwa et al. 2009, Chevalley 1975).

Lecithins and Polyglycerol Polyricinoleate (PGPR)

The emulsifiers most commonly used in the chocolate industry are the standard commercial lecithin and polyglycerol polyricinoleate (PGPR), which in small amounts can reduce the viscosity and the yield stress on the rheology of chocolate mass (Nieuwenhuyzen 2010). Although both act on rheological properties, the literature reports that commercial lecithin has greater influence on the plastic viscosity and the PGPR, greater influence in reducing the yield stress (Cunha, Quast, Luccas 2010, Afoakwa, Paterson, Fowler 2007).

Soybean lecithin is an important component in the structuring of chocolate, acting as an anti-bloom agent as well as in the plastic viscosity and yield stress of the chocolate mass. The commercial soybean lecithin can reduce by 10 times the amount of cocoa butter needed in the formulations, since this also reduces the viscosity of the mass, reducing overall production costs by saving cocoa butter (Schantz, Rohm 2005). Johansson and Bergenstahl (1992) reported that the lecithin acts on standard rheological properties and melting fat products, since it coats the sugar particles and at the same time modify the dynamic of fat crystallization, respectively.

The PGPR has a relatively small influence on the plastic viscosity (other than lecithin), however, a drastic effect on the yield stress. In literature, studies on rheology of chocolate use the combined effect of PGPR and soybean lecithin. Cunha, Quast, Luccas (2010) studied the influence of the addition of standard soybean lecithin and PGPR on the rheological properties of dark chocolate. The chocolate formulation was composed of 50% sugar, 40% cocoa liquor and 10% butter cocoa. The determinations of shear stress for calculating the Casson parameters were performed at temperatures of 40°C and 31°C, respectively, representing the start and end of the tempering process. The concentration of soybean lecithin added ranged from 0.3 to 1.4%. It was observed that the plastic Casson viscosity values at 40°C showed a decrease up to a concentration of 0.6% lecithin, remaining constant in the other concentrations, and the values obtained were between 6.5 and 2.4 Pa.s. The yield stress, measured at 40°C, also decreased in its value up to samples containing 0.6% of lecithin. After this concentration, it was observed a gradual increase. The values of this parameter to the range of 0.3 to 0.6% varied from about 57-35 Pa, and the range 0.7 to 1.4% ranged from 38.5 to 62 Pa. These values were extremely high compared to data presented in the literature.

The evaluation of rheological parameters at the beginning (40°C) and at the end (31°C) of the tempering process of the chocolate was performed varying concentrations of soybean lecithin from 0.3% to 0.8%. The results showed that the plastic viscosity values in the temperature of 31°C were slightly higher than the same samples at 40°C. At both temperatures, they showed a decreasing behavior as emulsifier concentration increased. For the yield stress measured at 31°C, the values were higher than those found at 40°C. However, it was found that there was a decrease in yield stress values to samples with concentration of 0.5% at both temperatures.

Samples containing 0.8% of emulsifier showed a very significant and smooth increase at temperatures of 31 and 40°C, respectively.

The significant increase in the values of the parameters at 31°C could be related to higher content of fat crystals present due to the pre-crystallization.

The authors have also studied the effect of adding 0.2% PGPR on samples containing 0.3%, 0.5% and 0.8% soybean lecithin for rheological evaluation. These tests were conducted at the end temperature of the tempering process: 31°C. They found that samples containing lecithin showed only decrease of viscosity, while those which also had PGPR showed an almost constant behavior and lower than those found with the samples containing only lecithin.

The yield stress values dramatically reduced in the samples containing PGPR, and the increase of lecithin content promote changes in the values of this parameter.

The same approach was performed by Stroppa et al. (2011). They studied the effect of different concentrations of lecithin and PGPR on rheological parameters (at 40°C) of dark chocolate (47% sugar, 43% cocoa liquor, 10% cocoa butter). In their study, soybean lecithin concentration levels ranged from 0.2% to 0.98% and from 0.1% to 0.58% for PGPR. The results showed that increasing the lecithin concentration promoted a reduction in the plastic viscosity, however, there was an increase in yield stress. The increase of the PGPR concentration promoted an increase in plastic viscosity values. Samples with higher concentrations of PGPR, which were composed of 0.2% and 0.5% lecithin and 0.5% and 0.58% of PGPR, respectively, showed a reduction of yield stress parameter to almost zero. The plastic viscosity values ranged from 1.89 to 5.94 Pa.s and the yield stress values from 0.03 to 24.95 Pa.

Phospholipids of lecithin may also influence the rheological behavior. By chemical modification, enzymatic or even solvent fractionation, the content of certain individual phospholipids may be modified. Bueschelberger (2004) noted that various phospholipids, such as PC and PE, have a different impact on the rheological properties. The PC is primarily responsible for reducing the viscosity, but with less effect on the yield stress. In contrast, lecithin rich in PE increases the ability to reduce the yield stress, but it is not very efficient in reducing the plastic viscosity. It should be emphasized that there are various factors influencing rheological behavior such as the distribution of particle size, type of sugar, fat thermal profile, etc.

The increased concentration of emulsifiers promotes increased rheological parameters, as seen by Stroppa et al. (2011), Cunha, Quast, Luccas (2010), Weyland and Hartel (2008) and Nebesny Zyzelewicz (2005).

Martin (1987) states that there is a limit point of emulsifier concentration on the mass, above which the mass consistency increases again. This increase may be related to steric stabilization problems or increased electrostatic repulsion between the emulsions formed by excess emulsifier (Wong 1995, Franco et al. 1988). Beckett (2008) reports that the addition of greater amount of lecithin promotes the formation of phospholipid bilayers surrounding the sugar. This situation may not only affect the rheology (Beckett 2008), but the crystallization and heat transfer processes. For example, thermal inertia, which exists in chocolate in the heating and cooling process, due to the large difference in thermal diffusivity between the particles of sugar and oil (Seguine 1991), could be enhanced, since the repulsion promote greater distance between the emulsion and, thus lower heat transfer. In addition, it has been observed in crystallization kinetic studies of fat addition in excess of emulsifiers delays the crystallization process (Cordiez, Grange, Mutaftschiev 1982). This may be due also to this steric hindrance or repulsion, thus hindering the packaging of the triacylglycerols.

However, it should be emphasized that each emulsifier has a different ability of stabilization. Afoakwa (2010) states that the PGPR can promote steric stabilization of sugar particles, reducing interactions, and thus affecting plastic viscosity and yield stress values in chocolates.

The evaluation of the mechanical resistance of chocolate is made by the breakdown voltage property that refers to the maximum breaking force applied to the center of the chocolate bars. The chocolate texture is determined by both the number and the size of the fat crystals. Numerous small crystals with a small amount of liquid fat promote chocolates with a hard texture. Also, the type of crystal and the ability to pack more cohesively, the presence of stable polymorphic forms will affect the chocolate texture. Chocolate with several large crystals tend to have a porous and unstable texture and may undergo breakage between these crystal conglomerates. It is therefore important during tempering to control the agitation, since it will impact on the size of the crystals (Lechter 2009). Svanberg et al. (2013) showed that when the chocolate presents the distribution of polymorphic forms more homogeneous in the mass, this would result in a chocolate with denser and more homogenous crystals than the opposite situation, resulting in a greater resistance to breakage.

The PGPR when added to the fat tends to form smaller crystal sizes than with the addition of other emulsifiers as set forth above (Bowser 2006, Rousseau et al. 2005). As small crystals is related to higher hardness, possibly the PGPR could induce this behavior.

Sorbitan Monostearate

Sorbitan monostearate was added to a formulation of dark chocolate for assessing possible modifications developed in the rheological parameters and texture of chocolates. This study was performed as an extension of the results showed by Masuchi et al. (2012 and 2014).

After evaluating the effects observed with the additions of different sorbitan monoesters (Masuchi et al. 2012, Masuchi et al. 2014), a formulation added with 0.5% of SMS on the oily phase (total amount of fat present in the formulation) was evaluated in the production of dark chocolate. Another dark chocolate formulation named as standard - manufactured without the addition of SMS - was also produced, and finally the two different formulations were evaluated for tempering temperatures, rheological parameters and ability to inhibit the formation of fat bloom during 90 days under two different storage conditions (constant temperature at 20°C and under 24 hours cycles of temperature changing between 20 and 32°C). The standard formulation of dark chocolate was made of sugar, cocoa powder, cocoa butter, lecithin (0.3% w/w) and PGPR (0.2% w/w), according to optimized concentrations presented by Stroppa (2011). The only difference between both formulations was the addition of 0.5% of SMS in one of the products.

The process conditions evaluated during the tempering steps of these two dark chocolate formulations for tempering temperature and reheating temperature were similar for the two formulations, and these values were respectively 29 and 30°C. The rheological properties of plastic viscosity and yield stress were determined using a digital rheometer (Brookfield, model TC500, US), with previously melted chocolate samples at 60°C and stabilized at a temperature of 40°C for measurements. The plastic viscosity (in Pa·s) and yield stress (in Pa) of the two formulations were, respectively, 2.8 and 11.8 for the standard formulation and 2.7 and 13.1 for the formulation added with SMS. It could be observed that the formulation containing sorbitan monostearate showed statistically significant lower plastic viscosity, fact experimentally established in the manufacturing processing when the formulation added with extra emulsifier (SMS addition besides lecithin and PGPR) was easier of molding.

As previously observed, the plastic viscosity decreased for the sample added with sorbitan monostearate compared to the standard formulation. In order to establish the possibility to reduce the lecithin content and/or PGPR when the mass already contains SMS, tests were made with dark chocolate containing only SMS and no addition of lecithin or PGPR.

After 23h of conching process, the chocolate mass only added with SMS emulsifier was evaluated as the need to add lecithin and also PGPR in the formulation. Thus, it was found that in the conditions evaluated in this study the addition of sorbitan monostearate as an unique emulsifier at a concentration of 0.5% (relative to the total fat formulation) can not replace the addition of lecithin and PGPR emulsifiers in the rheological characteristics for adequate molding and unmolding of the chocolate bars.

When a large amount of SMS (about 1.0% on total product mass) was added, there was a great reduction of the plastic viscosity, allowing the replacement of lecithin in this formulation, although PGPR still should be added for altering the yield stress.

As the cost of sorbitan monostearate is generally higher than lecithin, it was chosen in this case to proceed with the minimum use of SMS as a potential crystallization modifier (0.5% in the formulation lipid phase) and the addition of lecithin and PGPR at optimized concentrations of 0.3% and 0.2%, respectively. Thus, it was concluded that the emulsifier sorbitan monostearate acts as reducing slightly the plastic viscosity in chocolate, depending on its concentration in the formulation.

After evaluation of the rheological parameters of these two dark chocolate formulations (with and without SMS addition), both formulations were processed properly and a well-tempered dark chocolate was obtained for assessing the SMS ability to inhibit fat bloom formation.

Samples of each formulation were stored at constant temperature - 20°C - and under accelerated conditions for fat bloom formation - changing the temperature each 24 hours between 20 and 32°C. The chocolate bars were characterized by visual comparative ratings, snap test, whiteness index, SEM microstructure observations, DSC melting behavior and polymorphism by X-ray diffraction. Storage tests under normal and accelerated conditions confirmed a considerable deceleration in the fat bloom formation rate in samples containing the crystallization modifier agent, SMS.

The visual comparative ratings indicated that the dark chocolate samples stored in the isothermal condition of 20°C had no fat bloom visual development during the 90 days of storage, as expected. On the other hand, comparing both formulation stored at accelerated fat bloom conditions, the standard sample showed faster development of fat bloom than the sample with SMS addition, confirming that sorbitan monostearate can be a prominent emulsifier used as a anti-blooming agent for fat systems.

References

Afoakwa, E. O. *Chocolate Science and Technology*. 1st ed. United Kingdom: Wiley-Blackwell Publishers, 2010, p. 311.

Afoakwa, E. O., Paterson, A., Fowler, M. Factor influencing rheological and textural qualities in chocolate – a review. *Trends in Food Science and Technology*. v. 18, n. 6, pp. 290-298, 2007.

Afoakwa, E. O., Paterson, A., Fowler, M., Vieira. J. Comparison of rheological models for determining dark chocolate viscosity. *International Journal of Food Science and Technology*, v. 44, n. 1, pp. 162-167, 2009.

Akoh, C. C., Min, D. B. *Food Lipids: Chemistry, Nutrition and Biotechnology*. 3rd ed., Boca Raton: CRC Press, 2008. p. 928.

Aronhime, J. S., Sarig, S., Garti, N. Mechanistic considerations of polymorphic transformations of tristearin in the presence of emulsifiers. *Journal of American Oil Chemists' Society*. v. 64, n. 4, pp. 529-533, 1987.

Bastida-Rodríguez, J. The food additive Polyglycerol Polyricinoleate (E-476): Structure, applications, and production methods. *ISRN Chemical Engineering*, v. 13, pp. 1-21, 2013.

Beckett, S. T. *The Science of Chocolate*. 2nd ed., London: Royal Society of Chemistry Paperbacks, 2008, p. 96.

Beckett, S. T. *The Science of Chocolate*. London: Royal Society of Chemistry Paperbacks. 2000. p. 240.

Belitz, H. D., Grosch, W., Schieberle, P. *Food Chemistry*, Leipzig: Springer. 2009. p. 1070.

Bowser, A. Crystallization of cocoa butter. *The Manufacturing Confectioner*. v. 86, n. 9, pp. 115-118. 2006.

Bueschelberger, H. G. Lecithins. In: Whitehurst, R. J. (Ed.), *Emulsifiers in Food Technology*, 1st ed. Oxford: Blackwell Pub, 2004, pp. 1-39.

Cerdeira, M., Martini, S., Hartel, R. W., Herrera, M. L. Effect of sucrose ester addition on nucleation and growth behavior of milk fat-sunflower oil blends, *Journal of Agricultural and Food Chemistry*, v. 51, n. 22, pp. 6550-6557, 2003.

Chevalley, J. Rheology of chocolate. *Journal of Texture Studies*, v. 6, n. 1, pp. 177-196, 1975.

Co, E. D., Marangoni, A. G. Organogels: an alternative edible oil-structuring method. *Journal of the American Oil Chemists' Society*, v. 89, pp. 749-780, 2012.

Cordiez, J. P., Grange, G., Mutaftschiev, B. Solidification of stearic acid-water emulsion effect of the droplet-medium interface on nucleation kinetics. *Studies in Surface Science and Catalysis*. v. 10, n. 1, pp. 103-106, 1982.

Cunha, K. M., Quast, L. B., Luccas, V. Influência da adição de lecitina de soja e do Poliglicerol poliricinolato nas propriedades reológicas do chocolate. *Boletim do Centro de Pesquisa de Processamento de Alimentos*, v. 28, n. 2, pp. 321-330, 2010.

Cunha, K. M., Quast, L. B., Luccas, V. Influência da adição de lecitina de soja e do Poliglicerol polirricinoleato nas propriedades reológicas do chocolate. *Boletim do Centro de Pesquisa de Processamento de Alimentos*, v. 28, n. 2, pp. 321-330, 2010.

Dassanayake, L. S. K., Kodali, D. R., UENO S. Formation of oleogels based on edible lipid materials. *Current Opinion in Colloid and Interface Science*, v. 19, 432-439, 2011.

Deman, J. M. *Principles of Food Chemistry*. 3rd ed. Maryland: Aspen, 1999, p. 645.

Fennema, O. R. *Food Chemistry*. 3rd ed., New York: Marcel Dekker, 1996, p. 1067.

Franco, Z. A., Tavares, A. F. C. Gonzáles, G., Dibai, A. J. V., Porto, J. H. S. Emulsões e Borras (Parte 1): Teoria da formação e Quebra das emulsões, *Boletim Técnico Petrobras*, v. 31, n. 2, pp. 95-103, 1988.

Garti, N. Food emulsifiers: structure-reactivity relationships, design, and applications, In: Marangoni, A. G., Narine, S. *Physical Properties of Lipids*. New York: Marcel Dekker, 2002, pp. 265-386.

Garbolino, C., Bartoccini, M., Flöter, E. The influence of emulsifiers on the crystallisation behaviour of a palm oil-based blend. *European Journal of Lipid Science and Technology*, v. 107, n. 9, pp. 616-626, 2005.

Gunstone, F. D. Phospholipids. In: Gunstone, F. D. *Structured and Modified Lipids*. New York: Marcel Dekker, 2001, pp. 241-250.

Hasenhuettl, G. L. Overview of food emulsifiers. In: Hasenhuettl, G. L., Hartel, R. W. (Eds.). *Food Emulsifiers and their Applications*. 2nd ed., New York: Springer, 2008, p. 183.

Johansson, S., Bergenstahl, B. The influence of food emulsifiers on fat and sugar dispersions in oils. II. Rheology, colloidal forces. *Journal of the American Oil Chemists' Society*, v. 69, n. 8, pp. 718-727. 1992.

Killian, L. B. A., Coupland, J. N. Developing water-in-oil emulsions for chocolate products. *The Manufacturing Confectioner*, v. 92, n. 6, pp. 59-66, 2012.

Lechter, A. Tempering Chocolate - A Review. *The Manufacturing Confectioner*, v. 89, n. 1, pp. 53-58, 2009.

Lehninger, A. L., Nelson, D. L., Cox, M. *Princípios de Bioquímica*. 3rd ed. São Paulo: Sarvier, 2002, p. 975.

Liu, D., Ma, F. Soybean phospholipids. In: Krezhova, D. *For Enhancing the Diversity and Quality of Soybean Products*. Rijeka: InTech Press, 2011, pp. 483-500.

Lonchampt, P., Hartel, R. W. Fat bloom in chocolate and compound coatings. *European Journal of Science and Technology*, v. 106, n. 4, pp. 241-274, 2004.

Martin Jr, R. A. Chocolate. In: Chichester, C. O., Mrak, E. M., Schweigert, B. S. *Advances in Food Research.* San Diego: Academic Press, 1987, pp. 213-342.

Masuchi, M. H., Grimaldi, R., Kieckbusch, T. G. Effect of sorbitan monoesters addition on the crystallization and polymorphic behavior of cocoa butter. *Euro Fed Lipid*, Cracow, 2012. Avaiable: http://www.eurofedlipid.org/meetings/archive/cracow/5877/5877_0196.pdf. Accessed 20 July 2015.

Masuchi, M. H., Grimaldi, R., Kieckbusch, T. G. Effects of sorbitan monostearate and monooleate on the crystallization and consistency behaviors of cocoa butter. *Journal of the American Oil Chemists' Society*. v. 91, pp. 1111-1120, 2014.

McClements, D. J. *Food Emulsions: Principles, Practices, and Techniques*. 2nd ed. Boca Raton: CRC Press, 2005, p. 609.

Miskandar, M. S., Che Man, Y. B., Rahman, R. A., Aini, I. N., Yusoff, M. S. A. Effects of emulsifiers on crystallization properties of low-melting blends of palm oil and olein. *Journal of Food Lipids*. v. 13, n. 1, p. 57-72, 2006.

Miyasaki, E. K., Ming, C. C., Luccas, V., Kieckbusch, T. G. Influence of modified lecithins on crystallization behavior of cocoa butter. *Euro Fed Lipid*, Cracow, 2012. Available: http://www.eurofedlipid.org/meetings/archive/cracow/5877/5877_0194.pdf. Accessed 20 Jul. 2015.

Murdan, S., Gregoriadis, G., Florence, A. T. Novel sorbitan monostearate organogels. *Journal of Pharmaceutical Sciences*, v. 88, pp. 608-614, 1999.

Nasir, M. I., Bernards, M. A., Charpentier, P. A. Acetylation of soybean lecithin and identification of components for solubility in supercritical carbon dioxide. *Journal of Agricultural and Food Chemistry*, v. 55, n. 5, pp. 1961-1969, 2007.

Nebesny, E., Zyzelewicz, D. Effect of lecithin concentration on properties of sucrose-free chocolate masses sweetened with isomalt. *European Food Research Technology*, v. 220, n. 2, pp. 131-135, 2005.

Nieuwenhuyzen, W. V. Lecithin and other phospholipids. In: Kjellin, M., Johansson, I. *Surfactants from Renewable Resources*. 1st ed., United Kingdom: John Wiley and Sons, 2010, pp. 191-212.

Nieuwenhuyzen, W. V., Tomás, M. C. Update on vegetable lecithin and phospholipid Technologies. *European Journal of Lipid Technology*, v. 110, n. 5, pp. 472-486, 2008.

O'Brien, R. D. Fats and oils analysis. In: *Fats and Oils – Formulating and Processing for Applications*. Ed. O'Brien, R. D. 2nd ed. CRC Press, New York, 2004.

Oliveira, M. G., Ribeiro, A. P. B., Kieckbusch, T. G. Hard fats improve technological properties of palm oil for applications in fat-based products. *LWT - Food Science and Technology*, v. 63, 1155-1162, 2015 (a).

Oliveira, G. M., Stahl, M. A., Ribeiro, A. P. B., Grimaldi, R., Cardoso, L. P., Kieckbusch, T. G. Development of zero trans/low sat fat systems structured with sorbitan monostearate and fully hydrogenated canola oil. *European Journal of Lipid Science and Technology*. Online version, 2015 (b).

Peyronel, F., Marangoni, A. G. In search of confectionary fat blends stable to heat: Hydrogenated palm kernel oil stearin with sorbitan monostearate. *Food Research International*. v. 55, pp. 93-102, 2014.

Pokorný, J. Production, separation and modification of phospholipids for use in food. In: Gunstone, F. D. *Modifying Lipids for Use in Food*. Boca Raton: CRC Press. 2006. pp. 369-390.

Poletto, F. S., Oliveira, C. P., Wender, H., Regent, D., Teixeira, S. R., Guterres, S. S., Rossi Bergmann, B., Pohlmann, A. R. How sorbitan monostearate can increase drug-loading capacity of lipid-core polymeric nanocapsules. *Journal of Nanoscience and Nanotechnology*. v. 15, n. 1, pp. 827-837, 2014.

QuiñoneS-Muñoza, T., Gallegos-Infantea, J. A., Rocha-Guzmána, N. E., Ochoa-Martineza, L. A., Morales-Castroa, J., González-Laredoa, R. F., Medina-Torresb, L. Mixing and tempering effect on the rheological and particle size properties of dark chocolate coatings. *Cyta - Journal of Food*, v. 9, n. 2, pp. 109-113, 2011.

Raj, P. V., Nitesh, K., Chandrashekhar, H. R., Mallikarjuna-Rao, C., Venkata-Rao, J., Udupa, N. Effect of lecithin and silymarin on d-galactosamine induced toxicity in isolated hepatocytes and rats. *Indian Journal of Clinical Biochemistry*, v. 25, n. 2, pp. 169-174. 2010.

Ribeiro, A. P. B., Masuchi, M. H., Miyasaki, E. K., Domingues, M. A. F., Stroppa, V. L. Z., Oliveira, G. M., Kieckbusch, T. G., Crystallization modifiers in lipid systems. *Journal of Food Science and Technology*, v. 52, n. 7, pp. 3925-3946, 2015.

Rousseau, D., Hodge, S. M., Nickerson, M. T., Paulson, T. Regulating the $\beta' \rightarrow \beta$ polymorphic transition in food fats. *Journal of the American Oil Chemists' Society*, v. 82, n. 1, pp. 7-12, 2005.

Schantz, B., Rohm, H. Influence of lecithin-PGPR blends on the rheological properties of chocolate. *LWT - Food Science and Technology*, v. 38, n. 1, pp. 41-45, 2005.

Seguine, E. S. Tempering the inside story. *The Manufacturing Confectioner*, v. 71, n. 5, pp. 117-125, 1991.

Shah, D. K., Sagiri, S. S., Behera, B., Pal, K., Pramanik, K. Development of olive oil based organogels using sorbitan monopalmitate and sorbitan monostearate: A comparative study. *Journal of Applied Polymer Science*. v. 129, n. 2, pp. 793-805, 2013.

Smith, K. W., Bhaggan, K., Talbot, G., Malssen, K. F. V. Crystallization of fats: influence of minor components and additives. *Journal of the American Oil Chemists' Society*, v. 88, 1085-1101, 2011.

Stroppa, V. L. Z., Ribeiro, A. P. B., Luccas, V., Grimaldi, R., Gonçalves, L. A. G., Kieckbusch, T. G. Influence of soy lecithin and PGPR levels in chocolate crystallization behavior. 2011. *International Congress on Engineering and Food*. http://www.icef11.org/content/papers/epf/EPF 1081.pdf. Accessed 19 Jul 2015.

Svanberg, L., Ahrné, L., Lorén, N., Windhab, E. Effect of sugar, cocoa particles and lecithin on cocoa butter crystallization in seeded and non-seeded chocolate model systems. *Journal of Food Engineering*. v. 104, n. 1, pp. 70-80, 2011.

Svanberg, L., Ahrné, L., Lorén, N., Windhab, E. Impact of pre-crystallization process on structure and product properties in dark chocolate. *Journal of Food Engineering*, v. 114, n. 1, pp. 90-98, 2013.

Szuhaj, B. F. Lecithins. In: Shahidi, F. *Bailey's Industrial Oil and Fat Products*. New York: John Wiley and Sons. 2005, v. 3, pp. 361-456.

Tanno, H. Lecithin. *Ullmann's Encyclopedia Of Industrial Chemistry*. 6th Ed. Weinheim: Wiley-VCH Verlag GmbH and Co, DOI: 10.1002/ 14356007.a15_293, 2012.

Tisoncik, M. A. *Impact of emulsifiers on physical, sensory and microstrutural properties in formulated dark chocolate with an innovative educational approach*. 2010. p. 211. Dissertação (Mestrado em Ciência de Alimentos), Universidade de Illinois, Champaign, 2010.

Ushikubo, F. Y., Cunha, R. L. Stability mechanisms of liquid water-in-oil emulsions. *Food Hydrocolloids*, v. 34, n. 0, pp. 145-153, 2014.

Wang, F., Liu, Y., Jin, Q., Huang, J., Meng, Z., Wang, X. Kinetic analysis of isothermal crystallization in hydrogenated palm kernel stearin with emulsifier mixtures. *Food Research International*, v. 44, n. 9, pp. 3012-3025. 2011.

Weyland, M., Hartel, R. Emulsifiers in confectionery. In: Hasenhuettl, G. L., Hartel, R. W. *Food Emulsifiers and their Applications*. New York: Springer, 2008, pp. 285-304.

Wong, D. W. S. *Química de los Alimentos: Mecanismos y Teoría*. 1st ed., Zaragoza: Editora Acribia, 1995, p. 476.

Wu, Y., Wang, T. Soybean lecithin fractionation and functionality. *Journal of the American Oil Chemists' Society*. v. 80, n. 4. pp. 319-326, 2003.

Xu, Q., Nakajima, M., Liu, Z., Shiina, T. Soybean-based surfactants and their applications. In: Ng, T. B. *Soybean - Applications and Technology*. Rijeka: InTech Press, 2011, pp. 341-364.

Xu, Z. S., Wang, L. J., Liu, B. Z. Modification of soybean phospholipid and its key technique. *Cereals and Oils*, n. 11, pp. 3-9. 2008.

Yoshikawa, S., Kida, H., Sato, K. Promotional effects of new types of additives on fat crystallization. *Journal of Oleo Science*. v. 63, pp. 333-345, 2014.

BIBLIOGRAPHY

Cooking innovations: using hydrocolloids for thickening, gelling, and emulsification LCCN: 2013026186 Nussinovitch, A. Cooking innovations: using hydrocolloids for thickening, gelling, and emulsification / Amos Nussinovitch, Madoka Hirashima. Published/Produced: Boca Raton: Taylor & Francis/CRC Press, [2014] Description: xxxiii, 344 pages: color illustrations; 25 cm Links: Cover image http://images.tandf.co.uk/common/jackets/websmall/978143987/9781439875889.jpg ISBN: 9781439875889 (hardback) LC classification: TP453.C65 N87 2014 Using hydrocolloids for thickening, gelling, and emulsification Hirashima, Madoka. Summary: "Hydrocolloids are among the most commonly used ingredients in the food industry. They function as thickeners, gelling agents, texturizers, stabilizers, and emulsifiers. They also have applications in the areas of edible coatings and flavor release. Thanks to molecular gastronomy, they have now been brought to the forefront of modern cuisine -- available in small quantities for everyday use. While there a number of books devoted to production scale use of hydrocolloids, no book has yet addressed the needs of the chef. This volume is fully devoted to the fascinating topic of hydrocolloids and their unique applications in the kitchen.Each chapter addresses a particular hydrocolloid, protein hydrocolloid, or protein-polysaccharide complex. Starting with a brief description of the chemical and physical nature of the item, the authors go on to explore its manufacture and biological/toxicological properties. Emphasizing practical information for the professional chef and amateur cook alike, each chapter includes recipes demonstrating that particular product's unique abilities in cooking. Several formulations have been chosen specifically for food technologists, who will be able to manipulate the product for large-scale use or as a starting point for novel industrial formulations. "-- Provided by publisher. Subjects: Hydrocolloids. Cooking. Gums and resins. Stabilizing agents. TECHNOLOGY & ENGINEERING / Food Science. Form/Genre: Cookbooks. Notes: Includes bibliographical references and index. Includes bibliographical references and index.

Emulsifiers in food technology LCCN: 2014025604 Norn, Viggo. Emulsifiers in food technology / Viggo Norn. Edition: Second edition. Published/Produced: Chichester, West Sussex, UK; Hoboken, NJ: John Wiley & Sons Inc., [2014] Projected pub date: 1501 Description: pages cm ISBN: 9780470670637 (cloth) LC classification: TP455 .N67 2014 Contents: Introduction to food emulsifiers and colloidal system -- Lecithins -- Ammonium phosphatides -- Mono- and diglycerides -- Acid esters of mono- and diglycerides -- Di-acetyltartaric esters of monoglycerides (DATEM) and associated emulsifers in bread making -- Sucrose esters -- Polyglycerol esters -- PGPR polyglycerolpolyricinoleate E476 -- Propylene glycol fatty acid esters -- Stearoyl-2-lactylates and oleoyl lactylates -- Sorbitan esters and polysorbates -- Application of emulsifiers in dairy and ice cream products -- Regulation of food emulsifiers in the European Union -- Analysis of emulsifiers. Subjects: Food additives. Emulsions. Dispersing agents. Notes: Includes bibliographical references and index. Additional formats: Online version: Norn, Viggo. Emulsifiers in food technology Second edition. Chichester, West Sussex, UK; Hoboken, NJ: John Wiley & Sons Inc., [2014] 9781118921258 (DLC) 2014026673

Food additives data book LCCN: 2010043544 Food additives data book / edited by Jim Smith, Lily Hong-Shum. Edition: 2nd ed. Published/Created: Chichester,

West Sussex; Ames, Iowa: Wiley-Blackwell, 2011. Description: xvi, 1107 p.: ill.; 26 cm. ISBN: 9781405195430 (hardback) LC classification: TX553.A3 F562 2011 Smith, Jim, 1953- Hong-Shum, Lily. Summary: "The use of additives in food is dynamic, as consumers demand fewer additives in foods and governments review the list of additives approved and their permitted levels. Scientists also refine the knowledge of the risk assessment process and improve alternative additives, processes or ingredients. The revised second edition of this major reference work covers all the "must-have" technical data on food additives. Compiled by food industry experts with a proven track record of producing high quality reference work, this volume is the definitive resource for technologists using food additives"-- Provided by publisher. "The use of additives in food is a dynamic one, as consumers demand fewer additives in foods and as governments review the list of additives approved and their permitted levels. Scientists also refine the knowledge of the risk assessment process as well as improve analytical methods and the use of alternative additives, processes or ingredients. Since the first edition of the Food Additives Databook was published, there have been numerous changes due to these developments and some additives are no longer permitted, some have new permitted levels of use and new additives have been assessed and approved. The revised second edition of this major reference work covers all the "must-have" technical data on food additives. Compiled by food industry experts with a proven track record of producing high quality reference work, this volume is the definitive resource for technologists in small, medium and large companies, and for workers in research, government and academic institutions. Coverage is of Preservatives, Enzymes, Gases, Nutritive additives, Emulsifiers, Flour additives, Acidulants, Sequestrants, Antioxidants, Flavour enhancers, Colour, Sweeteners, Polysaccharides, Solvents. Entries include information on: Function and Applications, Safety issues, International legal issues, Alternatives, Synonyms, Molecular Formula and mass, Alternative forms, Appearance, Boiling, melting, and flash points, density, purity, water content, solubility, Synergists, Antagonists, and more with full and easy-to-follow-up references"-- Provided by publisher. Contents: Machine generated contents note: How to Use This Book.Part 1. Acidulants.Part 2. Antioxidants.Part 3. Colourings.Part 4. Emulsifiers.Part 5. Enzymes.Part 6. Flavour Enhancers.Part 7. Flour Additives.Part 8. Gases.Part 9. Nutritive Additives.Part 10. Polysaccharides.Part 11. Preservatives.Part 12. Sequestrants.Part 13. Solvents.Part 14. Sweeteners.Index. Subjects: Food additives--Handbooks, manuals, etc. Technology & Engineering / Food Science Notes: Includes bibliographical references and index.

How baking works: exploring the fundamentals of baking science LCCN: 2010006497 Figoni, Paula. How baking works: exploring the fundamentals of baking science / Paula Figoni. Edition: 3rd ed. Published/Created: Hoboken, N.J.: John Wiley & Sons, c2011. Description: xi, 516 p.: ill.; 28 cm. ISBN: 9780470392676 (pbk.) 0470392673 (pbk.) LC classification: TX763 .F54 2011 Summary: The essence of baking is chemistry, and anyone who wants to be a master pastry chem must understand the principles and science that make baking work. Learn the whys and hows of every chemical reaction, essential ingredient, and technique. Contents: Introduction to baking -- Heat transfer -- Overview of the baking process -- Sensory properties of food -- Wheat flour -- Variety grains and flours -- Gluten -- Sugar and other sweeteners -- Fats, oils, and emulsifiers -- Eggs and egg products -- Leavening agents -- Thickening and gelling agents -- Milk and milk products -- Nuts and seeds -- Cocoa and chocolate products -- Fruit and fruit products -- Natural and artificial flavorings -- Baking for health and wellness. Subjects: Baking. Notes: Includes bibliographical references and index.

Hydraulic fracturing chemicals and fluids technology LCCN: 2013655185 Fink, Johannes Karl, author. Hydraulic fracturing chemicals and fluids technology / Johannes Karl Fink. Edition: First edition. Published/Created: Amsterdam; Boston: Elsevier, GPP, Gulf Professional Publishing is an imprint of Elsevier, 2013. Description: xiii, 234 pages: illustrations; 23 cm ISBN: 9780124114913 0124114911 LC classification: TN871.27 .F56 2013 Summary: "Create hydraulic fracturing fluid formulations that meet project specific needs while protecting the environment and profitability." -- Cover, p.[4] "Demand for well stimulation chemicals and products continue to rise, and fracturing chemicals, when used properly, can protect the life of the well, the environment and engineers on the job. However, there are many challenges facing this frowing part of the industry, including lack of published references on chemical selection, basic mechanics of chemical components, environmental implications on chemicals used, and understanding the elements of hte products behind the trade names. Wel-known author, Johannes Fink, author of Petroleum engineer's guide to oil field chemicals and fluids, has published a quick look-up guide titled Hydraulic fracturing chemicals and fluids technology. Fink creates a concise and coprehensive reference handbook to enable the engineer to logically select and utilize the appropriate chemicals on any hydraulic fracturing job. The first book to be devoted entirely to hydraulic fracturing chemicals, fink eliminates the guesswork so the engineer can select the best chemicals needed on the job, while providing the best protection for the well, workers and environment."-- Cover, p.[4] Contents: General aspects -- Fluid types -- Thickeners -- Friction reducers -- Fluid loss additives -- Emulsifiers -- Demulsifiers -- Clay stabilization -- pH control additives -- Surfectants -- Scale inhibitors -- Foaming agents -- Defoamers -- Crosslinking agents -- Gel stabilizers -- Gel breakers -- Biocides -- Proppants -- Special compositions -- Environmental aspects -- Index. Subjects: Oil field chemicals. Hydraulic fracturing. Hydraulic fracturing--Environmental aspects. Oil wells--Hydraulic fracturing. Gas wells--Hydraulic fracturing. Notes: Includes bibliographical references and index.

Lubricants: introduction to properties and performance LCCN: 2014001449 Torbacke, Marika. Lubricants: introduction to properties and performance / Marika Torbacke, Åsa Kassman Rudolphi, Elisabet Kassfeldt.

Published/Produced: Chichester, West Sussex, United Kingdom; Hoboken, NJ: Wiley, 2014. Description: xvii, 191 pages: illustrations; 26 cm Links: Cover image http://catalogimages.wiley.com/images/db/jimages/9781118799741.jpg ISBN: 9781118799741 (hardback) LC classification: TJ1077 .T67 2014 Rudolphi, Asa Kassman. Kassfeldt, Elisabet. Summary: "Concise, accessible introduction to lubricants for engineers, technicians and researchers who are not experts in lubricant chemistry or tribology.Lubricants: Properties and Performance provides an easy to understand overview of tribology and lubricant chemistry, and bridges the gap between the two areas. The first part of the book is theoretical and provides an introduction to tribological contact, friction, wear and lubrication, as well as the basic concepts regarding properties and the most commonly made analyses on lubricants.Base fluids and their properties and common additives used in lubricants are also covered. The second part of the book is hands-on and introduces the reader to the actual formulations and the evaluation of their performance. Different applications and their corresponding lubricant formulations are considered and tribological test methods are discussed. Finally used oil characterisation and surface characterisation are covered which give the reader an introduction to different methods of characterising used oils and surfaces, respectively. Easy to understand overview of the properties and performance of lubricants Combines chemistry and tribology of lubricants into one unified approach Covers the fundamental theory, describing lubricant properties as well as base fluids and additives Contains practical information on the formulations of lubricants and evaluates their performance Considers applications of lubricants in hydraulics, gears and combustion engines"-- Provided by publisher. "Lubricants: Properties and Performance provides an easy to understand overview of tribology and lubricant chemistry, and bridges the gap between the two areas"-- Provided by publisher. Contents: Machine generated contents note: Preface xi List of Symbols xiii List of Tables xvii Part One LUBRICANT PROPERTIES 1 Introduction to Tribology 3 1.1 Tribological Contacts 5 1.1.1 Macroscale Contacts 6 1.1.2

Summary of Used Oil Analyses 146 References 148 8 Surface Characterization 149 8.1 Surface Characterization of Real Components 151 8.1.1 Examination of Nonused Surfaces 151 8.1.2 Examination of Used Surfaces 151 8.1.3 Characteristics of Application Examples 152 8.2 Microscopy Techniques 153 8.2.1 Visual Inspection 153 8.2.2 Light Optical Microscopy (LOM) 154 8.2.3 Optical Interference Microscopy 154 8.2.4 Atomic Force Microscopy (AFM) 154 8.2.5 Scanning Electron Microscopy (SEM) 155 8.2.6 Focused Ion Beam (FIB) 158 8.2.7 Transmission Electron Microscopy (TEM) 159 8.3 Surface Measurement 159 8.3.1 Statistical Surface Parameters 161 8.3.2 Contacting Stylus Profiler 162 8.3.3 Microscopy Techniques 163 8.4 Hardness Measurement 163 8.4.1 Macro and Micro Hardness 163 8.4.2 Nanoindentation 163 8.5 Surface Analysis Techniques 163 8.5.1 Selected Methods 164 8.5.2 Analysis Performance Parameters and Terminology 165 8.5.3 Depth Profiling and Chemical Mapping 167 8.5.4 Energy Dispersive X-Ray Spectroscopy (EDS) 169 8.5.5 Auger Electron Spectroscopy (AES) 170 8.5.6 X-Ray Photoelectron Spectroscopy (XPS) 173 8.5.7 Secondary Ion Mass Spectroscopy (SIMS) 176 8.5.8 Fourier Transform Infrared Spectroscopy 178 8.6 Summary of Surface Characterization Methods 179 8.6.1 Microscopy and Surface Measurement 179 8.6.2 Surface Analysis 179 References 182 Index 185 . Subjects: Lubrication and lubricants. TECHNOLOGY & ENGINEERING / Mechanical. Notes: Includes bibliographical references and index. Additional formats: Online version: Torbacke, Marika. Lubricants Chichester, West Sussex, United Kingdom; Hoboken, NJ: John Wiley & Sons Inc., 2014 9781118799703 (DLC) 2014011498.

Modernist cuisine: the art and science of cooking LCCN: 2011290050 Myhrvold, Nathan. Modernist cuisine: the art and science of cooking / Nathan Myhrvold with Chris Young and Maxime Bilet; photography by Ryan Matthew Smith and Nathan Myhrvold. Edition: 1st ed. Published/Created: Bellevue, Wash.: Cooking Lab, 2011. Description: 6 v.: ill.; 28-34 cm. ISBN: 9780982761007 (set) 0982761007 (set) LC classification: TX651 .M94 2011 Art and science of cooking

Young, Chris, food scientist. Bilet, Maxime. Smith, Ryan Matthew. Scope and content: An overview of the techniques of modern gastronomy. Nathan Myhrvold, Chris Young, and Maxime Bilet -- scientists, inventors, and accomplished cooks in their own right -- have created a six-volume 2,400 page set that reveals science-inspired techniques for preparing food. The authors and their 20 person team at The Cooking Lab have achieved new flavors and textures by using tools such as water baths, homogenizers, centrifuges, and ingredients such as hydrocolloids, emulsifiers, and enzymes. Contents: v. 1. History and fundamentals -- v. 2. Techniques and equipment -- v. 3. Animals and plants -- v. 4. Ingredients and preparations -- v. 5. Plated-dish recipes -- v. 6. Kitchen manual. Subjects: Cooking. Food. Gastronomy. Molecular gastronomy. Notes: Vols. 1-5 (34 cm.); v. 6 (28 cm.) is a spiral bound book. Includes bibliographical references, index, and glossary in v. 5 for all vols.

Top pops: 55 all-natural frozen treats to make at home LCCN: 2012014631 Zaiden, Emily. Top pops: 55 all-natural frozen treats to make at home / Emily Zaiden. Published/Created: New York: St. Martin's Griffin, 2012. Description: 111 p.: col ill.; 19 cm. Links: Cover image http://www.netread.com/jcusers2/bk1388/260/9781250004260/image/lgcover.9781250004260.jpg ISBN: 9781250004260 (pbk.) 9781250014627 (e-book) LC classification: TX795 .Z35 2012 Summary: "Ice-pops reinvented! Here are 50 delicious, all-natural popsicle recipes featuring such delectable flavorful combinations as pomegranate orange rose, rosemary grape, apricot honey yogurt, and cranberry clove. For both grown-ups and kids these treats are super simple and fun to make. All that is needed are readily-available ingredients and some basic kitchen equipment. The book includes pops that you can eat all year round and for every occasion, such as: [bullet] creamsicles and pudding pops that kids will love [bullet] fresh and fruity pops for hot summer days [bullet] coffee and tea pops for your caffeine fix [bullet] treats to serve during holidays or other celebrations [bullet] liquor-infused popsicles for grown-ups [bullet] healthy pops to help cure sore throats and upset stomachs...plus much more. A techniques section, ideas about which molds and sticks to use, plus charts to help you combine flavors and fruits inspires experimentation and insures perfect pops all year long. Championing The Popshop philosophy of eating local and organic, the recipes in this book are a great way to use up your farmer's market finds. There are no preservatives, emulsifiers, or dyes used to create these wonderful, 100% natural, guilt-free treats"-- Provided by publisher. Subjects: Ice pops. COOKING / Courses & Dishes / Desserts. Form/Genre: Cookbooks. Notes: Includes index.

INDEX

C

D

E

F

G

M

N

O

P

R

S

T

U

V

W

X

Y